ROBOTS AND ROBOTOLOGY

ROBOTS AND ROBOTOLOGY

by

R. H. WARRING

LUTTERWORTH PRESS · GUILDFORD, SURREY

First published 1983

ISBN 0 7188 2551 9

Printed and bound in Great Britain by
Ebenezer Baylis and Son Limited,
The Trinity Press, Worcester, and London.

Contents

List of Plates

(between pages 64 and 65)

The Battelle Micromouse

The 'Micromouse', shown on the front cover (and Plate 4), is an intelligent robot with a microcomputer 'brain' and an ability to work out how to traverse a maze after just two trial runs. On the third run it goes from start to finish without bumping into a wall, or making a wrong turn. In this respect it is more intelligent than human beings and robot designers are working on how this type of robot can be used in a more sophisticated way – perhaps domestic robots to vacuum carpets and even run household appliances.

The Micromouse was built by researchers at Battelle's Pacific Northwest Laboratories in the USA. Its grey glass fibre body houses about £100's worth of parts – but it took something like 500 man hours to assemble and 'debug' this super-rodent so that it could make 33 decisions each time it ran its 20-foot-square maze.

The mouse glides along on two main wheels driven by stepping motors – or motors which rotate the wheels an exact distance for each electrical pulse supplied them. The 'brain' counts the pulses to keep track of the distance covered.

Infra-red beams from light emitters on the underpart of the body are aimed at five sensors attached to arms extending from the upper body. The computer 'brain' stops the mouse when approaching walls or obstacles that interrupt the light beams.

On the first and second runs through the maze, the 'memory' capacity of the brain gathers data about the maze boundaries and identifies and enters the location of all obstructions. This is then 'processed' automatically to ensure an error-free run on the third attempt, because the 'brain' also has a capacity to work out how to respond under given conditions. For example, it signals 'left turn' if the mouse encounters a wall in front and a wall on the right. The 'brain', in other words, teaches itself the correct programme to follow.

Switch the battery off, though, and the brain's memory goes

blank. It is then ready to re-learn the course, or another obstacle course, in the next two runs.

Basically, although the Micromouse is really a 'scientific toy' (*see* Chapter 2), it is a true robot and one which effectively demonstrates the potential of intelligent robots.

Introduction

Before 1920 there were no robots. At least, the word *'robot'* had not yet been coined or 'invented', depending on how you look at it. And for the first three decades following, all robots were fictional beings — most of them threatening, frightening, malevolent characters. Not at all the willing slave of man the word robot was originally meant to mean. But fiction also produced his 'good' counterpart — the robot that under no circumstance would harm a human being. He even had a different name to emphasize that he was not an evil robot. He was called a *droid.*

'Real' robots that did useful work did not appear until the 1960s. They were the first of the *industrial robots* or 'metal collar workers'. Yet if you accept 'robot' as a general description for a mechanical, working replica of a living creature — man, animal or bird — real (that is, non-fictional) robots existed two thousand years ago, and in the 18th and 19th centuries became quite prolific. They were called then — and today correctly known as — *automata.* Our Victorian ancestors took great delight in these clever animated figures which could perform life-like actions, powered by clockwork. They have their counterpart today in *hobby robots* and *demonstration robots* which are primarily designed and built to entertain and amaze people, not perform useful tasks. Electric motors and electro-mechanical 'brains' have replaced clockwork mechanisms, miniature tape recorders have given them a voice, and microphones ears, and solid-state electronics the facility to command their actions by remote control.

Yet they are still not real robots — or are they? This question is difficult to answer. Popular conception is that a robot looks like a human being — a *humanoid*, in fact. Duplicating the form, limbs and actions of a human figure in mechanical form, to perform useful tasks, results in a robot which is clumsy, inefficient, and unstable. As soon as it lifts one foot off the ground to move, it is more likely to topple over than 'walk'. The simplest and most effective way of making a humanoid 'walk', in fact, is to put it on wheels. For that reason, industrial robots which perform working tasks, do not normally have mobility. They remain permanently at their work station and just reach out from there to perform their tasks.

Somewhat surprisingly in view of the statement made above, it is difficult to better the motions of the human arm for doing work from a stationary position. It can swing from side to side, move up and down, reach out and grasp. The majority of industrial robots, indeed, are based on similar joints, positions and movements. But the rest of the machine bears no resemblance to human form. It is mainly a support for the mechanical arm. Even the 'brain' which controls all the movements, may be located outside the robot's 'body'. They are weird looking devices which can continuously repeat a working sequence precisely and with exact timing, following a programme taught initially by a human operator. Equally, they will repeat any errors taught them by the human operator. As such, they have no intelligence. They are pure robots.

These are the 'first generation' industrial robots, now working in thousands in factories in the more advanced countries throughout the world. Robots with more advanced 'brains', giving them the ability to sense, learn and adjust by their own experience, are only just beginning to appear. These 'second generation' robots possess a marked degree of 'intelligence' located in an advanced computer-type brain. Their mechanical side is very little different. They merely have more 'brain power', giving them a certain independence of action over what they have been taught (or programmed to do) by a human operator. At the same time this extra ability has also to be taught or 'written into' the brain by a human operator, communicating with the robot by machine language or *logic*. But the one thing an electronic brain cannot be programmed with is *emotion*. It can only analyse, separate, compute and react from information it receives on a logical basis or 'yes/no' reaction. It is this that distinguishes the robot from a human being, however advanced it may be.

Second generation robots are still only in the development stage — although some of the present first generation robots are clever enough almost to qualify for this description. Third generation robots will inevitably follow, and here the prospects are somewhat awesome. It is mainly a matter of brain volume or number of brain cells. Micro-electronics has reduced the size of a computer, with similar brain volume, from that of a large room to little larger than a cigarette packet. Even so, to produce a computer brain with the same number of cells as a human brain would probably need a volume larger than a small factory – and take years to programme. But given such a computer brain – and further developments in micro-technology could shrink it in size – theoretically, at least, the third generation robot could be given all the intelligence of a human being, with an infallible memory – and possibly (although this is highly questionable) even emotional reactions.

This conception of the robot of the future holds potential danger. Given such power it could become self-interested, self-preserving, self-generating – a reversion to the malevolent robot form originated by fiction. Instead of the man-made slave it could become the master of man, and ultimately would have no need of man at all. Fortunately this is an unlikely possibility. What is far more likely is the deliberate development of malevolent robots *by* man to further his own ends. In fact, this type of man-made 'brain' already exists in the command systems of modern missiles which seek out and lock on to a target.

THE GROWING ROBOT POPULATION

In 1960 there were probably not more than a few hundred true robots in the world. Since then the robot population has grown at a remarkable rate, reaching, by the end of 1981, an estimated 100,000 or more. Discounting fictional robots and 'scientific toy', and demonstration types (which probably account for a few hundred only in total), all were *industrial robots*. And it is estimated that the industrial robot population will continue to expand at a rate of some 35 per cent per year through the 1980s.

A current (1982) estimate of first-generation industrial robots (*see also* Chapter 3) installed in factories throughout the world is:

Japan	7–8000
USA	4000
Germany	1500
Sweden	1350
UK	400
Rest of Europe	200

The remainder of the current robot population is made up of the simpler type of 'pick-and-place' or 'low technology' robots.

No figures are available for Russia, although they are thought to have more than the USA, putting them in second place. (Certainly they were ahead of the USA in 1980). Japan leads the world in robot manufacture and application, and has done so from the very first. The accompanying graph shows the growth of Japanese robot production, and estimated figures through to 1990.

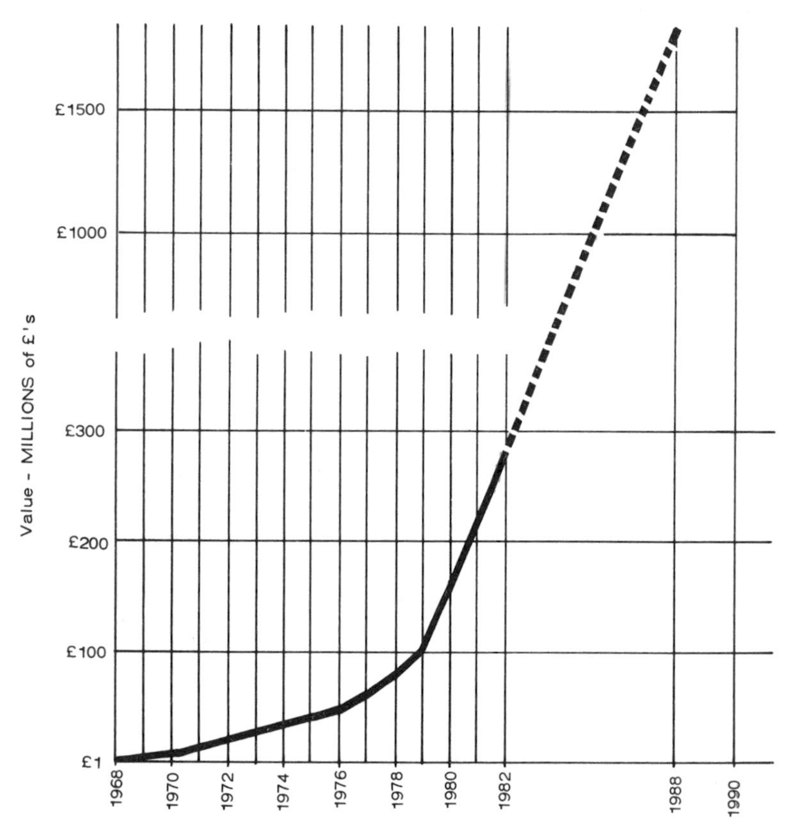

Japanese industrial robot production by value
(source *Japanese Industrial Robot Industry Association)*

Certainly, too, Japan had hundreds, if not thousands, of simpler industrial robots installed and operating before 1968. Seiko, for example, developed its own mini-robot system some thirty years ago specifically to assemble a wide range of mechanical watch movements. These were no longer required when quartz electronic movements took over from mechanical movements — yes, robots too, can be made redundant! But Seiko have continued to use the experience gained to combine to produce 'pick-and-place' robots (amongst other types) capable of a constant placement accuracy of the order of five ten-thousandths of an inch (0.0005 in.).

CHAPTER 1

Defining Robots

The word *robot* is an invention of fiction. It was coined from the Czech word 'robotnik', meaning a slave or serf and used by Karel Capek in his play *R.U.R.* produced in 1920. *R.U.R.* stood for *Rossum's Universal Robots*, which were the subject of the play — mechanical machines modelled on the lines of human beings with an immense capacity for work and none of the human weaknesses. At least, that was the original idea behind them. In the play these willing mechanical workers became used for war, and eventually turned against their human masters.

The first robots — fictional as they were — were thus 'evil' *androids* (human-looking machines). It took another writer, Isaac Asimov, to restore the balance, as it were, by proposing the Three Laws of Robotics to which all robots should conform. These were:

1. A robot must not injure a human being, or, through inaction on its part, allow a human being to come to harm.
2. A robot must always obey orders given to it by a human being, except where these conflict with Law 1.
3. A robot must protect its own existence, but only as long as this does not conflict with Law 1 and Law 2.

This defines the 'good' robot or *droid* of fiction; but equally summarizes the desirable features of any type of robot — fictional or real.

In the meantime, the description 'robot' has become a generic term covering a variety of different robot types which can be given individual names. But all are essentially man-made machines — in fiction or fact. Out of the latter has grown a new technology, *robotology*, or the means by which these machines are put together and made to work. An alternative description for robot technology is *robotics,* although this is more correctly descriptive of the pure science of robotology.

The definition of *robot* can now be expanded on the following lines:

ROBOT: strictly speaking the general name for unit or machine which is able to perform human-type actions and functions, without necessarily having a human appearance. In the broadest sense, the description robot is also applied to automated machines, particularly those capable of carrying out mechanical functions which can also be performed by humans (with or without the use of tools). By the first definition, too, brainpower is also a human function, so an electronic brain can also be considered as coming into the category of robotology. For example, a computer is an electronic counterpart of (part of) a human brain.

ANDROID: robots built to look like human beings and perform some of the actions, etc., carried out by humans. For example, it may 'talk'. It does not necessarily have to duplicate all the basic movements of a human being.

On this basis two classes of *androids* could be defined:

1. Robots that *do* have legs, arms, a body and head and duplicate the human body in this respect – e.g. *see* Fig. 1.1. They are basically a modern type of *automata,* probably more readily described as *realistic robots.*
2. *Functional androids* which have incomplete resemblance to humans. For example, they may achieve mobility through wheels rather than legs since this provides better functional performance.

HUMANOID: the popular concept of a robot. It has human appearance (symbolic or actual) and mechanical capabilities to perform various tasks. It is difficult to decide when an android is better described as a humanoid, and vice versa.

AUTOMATON: Basically a 'scale model' of a human being (or even an animal, bird, etc.) with the emphasis on realistic appearance, together with automated working features. It is not a true robot as such in the modern sense.

CYBORG: a CYBernetic ORGanism, the definition of which is now applied in three different ways. Originally it meant a human being modified by the addition of artificial limbs or organs (e.g. the *Six Million Dollar Man* and *The Bionic Woman*). The complement of this is another fictional type — the robot with a human brain, properly called a *cyberman*. The definition of a practical (as opposed to fictional) cyborg is a biological type of robot with some artificial intelligence or thinking capability.

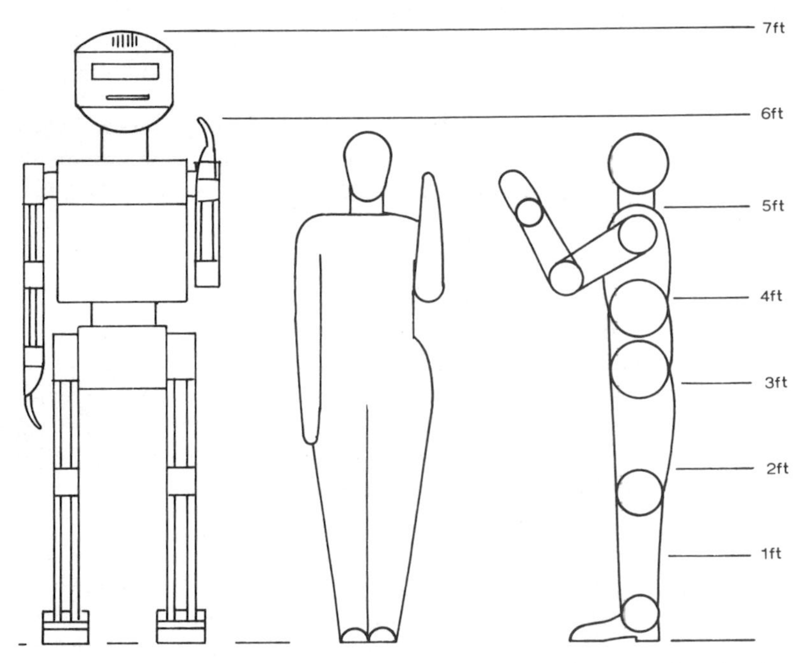

1.1 Proportions of typical human figure compared with 7 ft tall android.

CYBOT:a mechanical robot with the same 'mental' capabilities as a (true) cyborg.

CYBERT: the ultimate development of the *cybot*, as presently visualized, combining the characteristics of a humanoid with full mechanical abilities, intelligence and — most significant of all — an ability to learn by its own experience.

DROID: a true slave robot which obeys orders, may not injure or allow a human being to come to any harm, and puts its own self-protection behind these first two requirements. At present these are fictional, i.e. 'good' robots totally devoted to their human masters. But the stated premises are a desirable feature in non-fictional robots as well.

INDUSTRIAL ROBOT: this description is given specifically to machines and devices designed to both manipulate and transport parts, tools, implements, etc, through a programmed manufacturing task. A machine to take the place of men on a production line, or as it has been aptly called a 'metal collar worker'. A purely *functional* robot combining four main systems — the mechanical structure, power drive, a sensor and servo system, and a functional control system.

MANIPULATOR: is the simplest form of industrial robot, worked by an operator. In fact all industrial robots are manipulators, but all other types which are programmable come under the heading true robots.

AUTOMATA OR ROBOTS?

Under this heading we come up against a real problem in definitions. Is any machine which is a working model of a life form a robot? If so, mechanical toys are robots although more correctly called *automata* — and they pre-date the use and conception of 'robot' by some two thousand years. There are records of statues that drank water and birds that sang being made in the years 300–200 BC, followed by numerous other automata powered by water power and simple mechanical systems.

After the invention of clockwork mechanism in the eighteenth century, automata became more and more complex and intricate, mostly based on doll-size human figures capable of performing life-like tasks. The highly realistic clockwork doll produced by Jaquet Droz around 1775, for example, could play a model piano, whilst another could write words.

Automata were produced in quantity right into the early part of the present century. Similar principles have been revised in modern dolls, toys and working models, but using modern techniques, e.g. electronics for voice production, or even remote control. None is available as robots as we understand the term, although they may perform certain robotic functions.

Trying to separate automata from robots becomes even more difficult if we accept a working human figure as being a robot. Although, by robotic definition it is a *humanoid*, this would be called a robot by most people if 'he' actually walked freely. George Moore built such a 'robot' in 1893. It was powered by a simple steam engine housed within the body, with steam exhausting through a hollow cigar in the figure's mouth.

'Hobby' Robots

Many ingenious individuals have constructed walking, moving, talking robots of humanoid type – life-size and sometimes larger – powered by clockwork or electric motors. Numerous robots of this type have been built from Meccano, others have been built from scratch. They could all be categorized as 'hobby robots' – made to amuse, entertain or amaze onlookers wherever they are demonstrated. Some could even be described as stringless puppets.

Robots of this type have also been used commercially, for example, to hand out leaflets and 'talk' to potential customers at an exhibition. Others have been developed by firms for publicity purposes, or as first prototypes of possible 'domestic' robots. More properly these could be called *demonstration* robots. All have one feature in common: they are primarily designed as *humanoids*, to be instantly recognizable as the popular conception of what a robot should look like. This inevitably limits the functions possible to perform via true robotics, i.e. many – and particularly the hobby robots – tend to become a mixture of robot and automaton.

One basic problem in trying to construct a true humanoid is the sheer number of muscles and movements in the human body, and the vast number of brain cells needed to control them properly. The human body, for example, has some three hundred different muscles capable of producing some five hundred rudimental motions alone. That could call for 300 electric motors operating 500 different mechanisms, many interconnected – and a control system to work everything in correct sequence or synchronization.

Demonstration Robots

Several leading companies in the field of robotics have produced demonstration 'robots' for publicity purposes. *Elektro,* for example, was made by Westinghouse some thirteen years ago as a 'popular' example of how a future humanoid may look and behave. *Elektro* was a 7-foot tall humanoid who could speak some eighty different words, count numbers, walk, talk, salute and perform a total of 26 different actions (including distinguishing different colours). 'He' was powered by eleven electric motors, had a brain consisting of 82 relays, and weighed 260 pounds. ('He' also had size 18 feet!) *Elektro* responded to commands spoken into a microphone.

Today *Elektro* is completely outclassed in performance and technology. It is clumsy, has very limited functions and is completely outdated in using relays instead of solid-state electronics for a 'brain'.

Klater, produced by Quasar Industries a few years ago, is a much more enlightened *demonstration robot*. 'He' travels on wheels instead of legs, has 'his' conical body dressed in human-type clothes and has an electronic brain. *Klater* is a sort of prototype *domestic robot* capable of answering the door, greeting guests and leading guests into the house, directing them where to be seated, explaining that 'he' will now go out and find his master and telling him that they have arrived (even repeat any message the guests may have given him in the meantime). To keep *Klater* company he was later joined by a female humanoid with a most appealing face and dress. Like *Klater*, however, 'she' was also a compromise between 'mechanics' and 'appearance' in having a cone shaped body and wheels instead of feet. 'She' was a lot prettier than *Klater* since her skirt hid the absence of 'legs'.

Art Robots

The description 'art robot' is a new one, applied to advanced automata or humanoids used in tableaux – either for advertisement or entertainment. There are the tableaux designed for department stores, for examples, where the models move, walk, talk or otherwise perform human-like actions. Tableaux for entertainment may range from a sort of animated Madame Tussaud's to complete plays performed by robot figures.

Art robots, unlike domestic robots, already exist and are made by several manufacturers. Leaders, at present at least, are the Japanese productions of Mizuro of Tokyo, who only allows his robots to be hired out by those who will use them without attempting to discover the secrets of their construction. Recent humanoids completed by him include life-like robots of President Kennedy and Marilyn Monroe.

Simulators

The use of simulators is well known and well established as a method of training pilots as a supplement to actual time-in-aircraft training. This has considerable advantages both in cost savings and in giving the pilot experience of handling potentially dangerous situations far beyond those which can be demonstrated in normal flight training.

These — and other types of simulators used in operator training — are not robots. They are simply machines which can be programmed to reproduce occurrences on a visual and/or audio, and/or physical basis — often all three.

Once the 'machine' becomes a humanoid, however, it has the appearance of a true robot. An example is a *body simulator* in a life-like, life-size model of a human body with built-in movements and reactions, again designed for training in techniques, e.g. in resuscitation. The 'body' is apparently dead when the student starts first aid. If the resuscitation technique applied is correct, the 'body' recommences breathing, heartbeats recommence, the eyelids start to flutter and the pupils react to light or the 'body' comes back to life. Built-in recorders can at the same time record the effectiveness and proficiency of the student carrying out resuscitation.

There are not many body simulators in use today but their numbers will undoubtedly continue to grow in both student use and (with a greater degree of sophistication) student-doctor training. They are another form of state-of-the-art development of humanoids — actually in use and not just visualized as a future possibility. Whether they are true robots or not is questionable.

CHAPTER 2

Popular Robots

The most familiar robots are fictional characters, known to millions of people all over the world through their appearance in films or on the TV screen. They are, of course, not robots at all. At best they could be called a form of automata (and some are not even that), dressed up in the modern science-fiction image. Their capabilities are limited only by the limits of the imagination of their creators. Yet technology has a habit of catching up on and overtaking science fiction (like space travel) – so who knows when real robots *will* outperform their fictional counterparts?

Frankenstein's monster is one of the earliest fictional robots and, since it (he?) had the form of a human being, could be called an *android* (although the definition of android concerns a robot built from artificial material, not from parts of dead bodies!). In fact the *Golem* was a true (fictional) android in the sense that he was a man of clay, brought to life by a magic formula. The *Golem* predated Frankenstein's monster in appearance before the public in a film of the same name made in 1914. Frankenstein's monster did not appear on the screen until 1931. Between these two dates many robot films were produced with 'Metallic men' robots, most of whom seem to have ended up by going berserk, finally disintegrating into a miscellany of gear wheels, springs and wires.

There is a very good reason why many earlier film makers preferred androids and humanoids to other types of robots. Their parts can be played by human actors without requiring special effects, thus making the film cheaper and easier to produce. This still holds true today, but probably more for limited-budget TV productions than films.

The 1950s onwards saw further developments in robots appearing in films – the alien form of android in *Target Earth* (1954) and *Gog* (1954). The two non-humanoids in the latter film, *Gog* and *Magog*, were malevolent robots.

You've seen them on TV

The malevolent form of robot was also the main feature of the TV series *Dr. Who* (1974/75), which also included a cyberman or two for good measure, in the form of the Daleks (distinctly non-humanoid!); and in the malevolent android Maximilian in the film *Black Hole* (1980).

At the same time a number of likeable robots appeared in films, notably C-3PO and R2D2 in *Star Wars* (1977). C-3PO was the 'protocol robot' (a humanoid), and his smaller, tubbier companion the 'maintenance robot' (a non-humanoid with a body the shape and size of a dustbin with a hemispherical top). Devoted to the service of man, they were true *droids* — a robot type that exists only in fiction. In fact they were merely costumes with actors inside.

Hobby Robots

Hobby robots, as defined in the previous chapter, are interesting mainly because they 'work' and (usually) have humanoid features. One can only admire the ingenuity of their constructors in the way of producing something different in working models — if one accepts that a 'model' is not necessarily a miniature but can also be larger than life-size.

Certainly all the earlier robots of this type were self-contained as it were, operating movement by a system of mechanical linkages, levers, cams, etc. In other words they were mechanically pre-programmed. With the advent of practical radio-control systems for models, the hobby robot was given more scope. Movements no longer needed to be initiated by a built-in programmer. They could be worked by servos with remote control by a human operator, i.e. controlled by human intelligence. Thus a relatively simple mechanical hobby robot — they are ingenious rather than complex machines — no longer had to 'walk' into objects it could not sense, or follow only a limited pre-programmed action schedule. Its performance is limited only by the extent of its functional capabilities, not a set of repetitive actions like automation. It 'borrows' the intelligence of its human controller, as far as this can be utilized by its own mechanics.

Radio control of robots is, in fact, a simple and very much cheaper alternative to an electronic brain — and a logical choice for a hobby type robot. Whether it qualifies as a true robot, however, is debatable. More realistically it would be described as a radio-controlled automaton.

A similar principle is, of course, used with industrial robots of the simple *manipulator* type. These are essentially mechanical systems which are *directly* operator-controlled from a point remote from the working area of the manipulator. Remote control in this case is powered by mechanical linkages or electrical signals controlling the operating mechanisms. It is a much cheaper type than a programmed robot.

At the same time *directly* linked remote control, as opposed to remote control by radio signals, has the advantage that 'feel' can be built into the system, greatly assisting the operator in making the manipulator undertake the task demanded. With 'feel' present — and plenty of practice on the part of the operator — the otherwise clumsy mechanical manipulator can be made to perform quite intricate and delicate jobs.

The significance of 'feel' and means by which this faculty can be provided is described in some detail in Chapter 8. It could be highly desirable, even in hobby robots (if they are designed to shake hands with someone, for example).

SCIENTIFIC 'TOYS'

Certain robots made by individuals, or companies and designed to demonstrate artificial 'intelligence' are probably best classified as scientific 'toys'. They are basically demonstration type robots but are not merely pre-programmed (or remotely controlled by radio) but incorporate some intelligence or ability to learn and act accordingly on their own.

An early example was Greg Waller's tortoise, constructed some forty years ago in the form of a small wheeled robot with a body shell in the shape of an inverted oval bowl with a device looking like a periscope sticking out of the top. This was, in fact, a photo-electric eye or light sensor. Driven by a battery-powered electric motor this little robot had the ability to move around freely over any flat floor, sensing objects in its way through its eye and avoiding collision with them. At the point where the battery showed signs of running down a new sense was signalled in the robot's 'brain'. It started to search for the light in its hutch (previously it had ignored this and simply ran about freely). Having determined where the hutch was, it ran up to it, entered and plugged itself in to recharge its batteries ready for another run.

Basically only the search-and-find facility was robot intelligence. It had to find its own way from where it was when its battery started to run down. The need to start this search was a pre-programmed feature — merely a relay which switched from one position to another when the battery voltage reached a predetermined low level.

Other examples of scientific 'toys' include *electronic mice* with varying degrees of 'intelligence' to seek their way through a maze; and the robot built by the Stanford Research Institute in the USA which in a small room could work out how to locate and shift an inclined ramp into such a position that it could climb up it on to a platform placed anywhere in that room.

Clever as they were, the most these — and other — scientific 'toys' really demonstrated was the complexity involved in imparting even a miniscule fraction of human intelligence into a robot brain. They had no direct practical application as such, hence 'toy' is really justified in their description. It is far easier to pre-programme a robot than to provide it with a 'brain' to work out its own problems, as first-generation industrial robots have proved. The most advanced of these — the playback type — simply have a memory, not intelligence, to repeat what they are taught by a human operator.

CHAPTER 3

Industrial Robots

It is difficult to say when the first industrial robots were put to work in factories. It depends on what you accept *are* robots, as distinct from *automation*. Automation, or automatic control of machines, started in the 1950s and became firmly established in the 1960s. The programmable robot as a separate entity, i.e. not a control system built into the machine itself, only began to appear in any numbers in the late '60s. Since then the growth of robots has been phenomenal in Japan, Russia and the USA (in that order, incidentally). British industry has been reasonably slow in adopting industrial robots, so much so that as far as original design is concerned we have missed out almost entirely on 'first generation' robots. Ironically, the first computer-controlled steel rolling mill which established a world lead in *automation* in the '60s was British.

'First generation' robots, in the generally accepted sense, are manipulators which can be programmed to perform particular operations. These fall into four different categories:

(i) Fixed sequence robots.
(ii) Variable sequence robots.
(iii) Numerical control (NC) robots.
(iv) Playback robots.

These are the types currently in use in industries in developed countries, accounting for a world robot population of about 15,000 in 1981 in Japan, USA, Sweden, Germany, Italy, the UK and the rest of Europe. Types (i), (ii) and (iii) may also be referred to as *low technology* robots, unless allied to a computer 'brain'. Type (iv) may be described as a *high technology* robot.

'Second generation' industrial robots have yet to be developed in any numbers. This description is given to a robot with sensory perception — e.g. visual or tactile, or both — which can detect changes *by itself* in the work condition or work environment, make a *decision* on what action to take to apply any correction or adjustment necessary, and proceed with its operation.

Strictly speaking, these are 'third generation' robots. Manual manipulators were the real 'first generation' robots; and programmable robots the 'second generation' robots. Current robotic terminology, however, excludes manual manipulators from the category of true industrial robots and starts with programmable manipulators (robots) as the first generation. The following descriptions then cover these types.

Fixed Sequence Robots

Fixed sequence robots are machines or devices capable of performing successive steps of a given operation in a predetermined sequence, condition and position. The preset information is normally built into the device and cannot readily be changed. That is, the robot is designed for a specific operation. A simple pick-and-place robot with a rectilinear mechanical system is a typical example, but many more complex mechanisms may be of this type.

3.1 Typical layout of an assembly line combining robots with manual work stations.

Variable Sequence Robots

This type is essentially the same as a fixed sequence robot except that the set information can easily be changed. Again, taking the simple case of a pick-and-place robot, the required programme may be preset by plugging various electrical connections into a plug board. To change the programme, the positions of the plugs are rearranged.

It follows that the original design of the mechanical system is more complex than that of a fixed sequence robot. It has to be capable of providing a variety of different movements or movement ranges to accommodate different programmes. Usually, too, in the present day variable sequence robot the plugboard is replaced by a microcomputer.

Numerical Control Robot

This type of robot has the same mode of control as a numerical control machine. Sequence, condition and position is commanded by numerical data fed into the machine. These programmes (software) may be in the form of punched tapes, punched cards or digital switches. Like the variable sequence robot, it is designed to accommodate a range of different programmes.

Playback Robot

This is one of the most interesting types of robots. It incorporates a memory but has to be *taught* the sequence, positions and operations required by a human operator. Having once been taught these, it holds the information in its memory. When needed, this information is recalled (played back) and the operations are repetitively executed automatically from memory.

The block diagram of Fig. 3.2 shows basically how this system works. The human operator has a control unit to work the robot as a manual manipulator. He moves the manipulator to the required first position and stops it there. Pressing a *second* button then impresses a signal representing this position in the *memory*. The operator then moves the manipulator to the next condition or position required, and so on, pressing the *second* button at each sequence position, until the full programme has been recorded in the memory.

On some playback robots there is no separate manipulator to be plugged into the robot for manual instruction. Instead the operator grasps and manipulates the robot arm directly, marking each stop position in the memory by pressing a button on or near the grip. The principle involved is exactly the same.

The robot, in fact, has been 'taught' its programme step-by-step. From then on the human operator is no longer required. It will

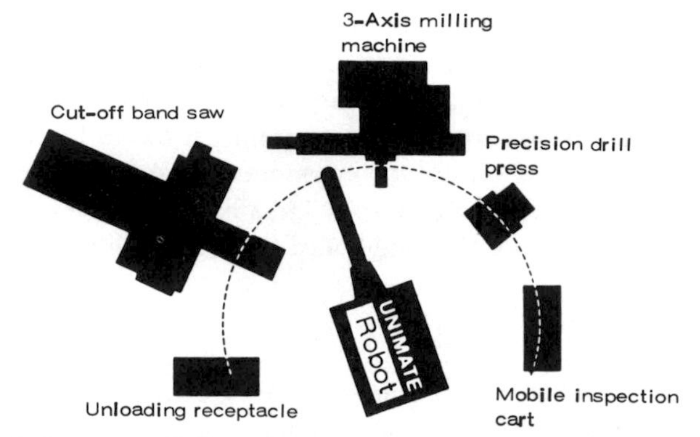

3.2 Robots are widely used for automating small-batch operations.

repeat its working programme when activated from its own memory. Equally the original programme can be cancelled at any time, i.e. removed from the memory, and replaced with a different programme.

A particular advantage of this type of robot is that only the 'start' and 'stop' positions are recorded in the memory. Time spent by the human operator in getting the positions exactly right does not come into it. He can take as long as necessary, pressing the 'memory' button only when the exact positions have been set. When the robot is operating from its own memory it then moves from start to stop positions at a constant speed.

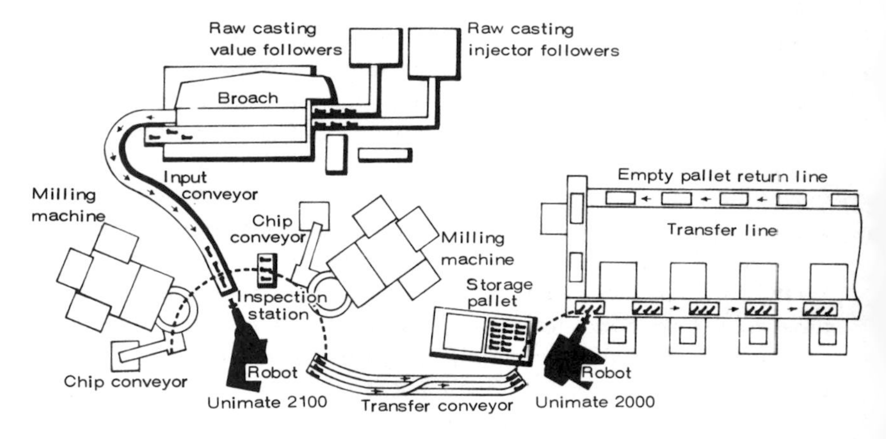

3.3 On more complex assembly lines 'pick-and-place' robots operate at various strategic positions.

In practice the human operator may continually re-adjust the programme through trial runs of the programme; or even over weeks of operation in order to get the most effective programme. This can apply particularly if several robots have to be programmed to work together on the same job. Only trial-and-error adjustment can get them all working efficiently in a co-ordinated manner, e.g. one robot not getting in the way of another; and all finishing at the same time. Engineers have a name for this extended teaching programme — *robot choreography*.

Point-to-Point and Continuous-Path

The form of teaching just described is called point-to-point. The robot is told to move from point to point on 'start' and 'stop' basis, and perform some function at each 'stop' point. This system is well suited to certain operations, like spot welding, where a jerky movement followed by a pause is acceptable. Other operations, like spray painting, for example, require a smooth movement following a continuous path.

In this case the robot requires a rather more sophisticated memory so that it can *continuously* record *all* the movements initiated by the human operator teaching it. Equally it will follow and repeat all the mistakes made by the human operator. Teaching a playback robot a *continuous-path* programme thus demands a highly skilled operator in the first place. Again, in the light of operating experience, this programme may have to be re-taught to get the best results.

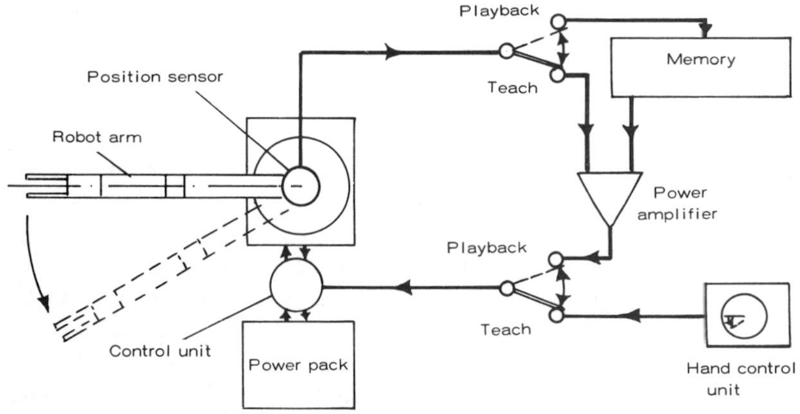

3.4 Block diagram of playback robot system.

FOR the INDUSTRIAL ROBOT

1. It takes the boredom out of routine, repetitive functions, releasing the operatives concerned for more interesting work.
2. They work with repeated precision and accuracy not always possible with human operators.
3. They maintain a constant quality of productive work; they do not get tired.
4. They provide continuity of production; they do not require rest periods or tea-breaks.
5. They can safely handle hot, dangerous or toxic components without fear of contamination or danger to human operators.
6. They add to job security by contributing to the consistency of the product.

AGAINST the INDUSTRIAL ROBOT

1. The general unwillingness of the labour force in most countries to accept robots and likely redundancy as a consequence.
2. Items 1 and 6 in the 'FORS' are influenced by national and international economic conditions and politics. They are not necessarily true at any one time.
3. Industrial robots require large capital investment. Debatable because capital used is not a true cost. It has to be considered with returns from robot 'labour' and the resulting pay-back period (i.e. when the robot has paid for itself in increased production and various savings). Experience in many industries has shown that the pay-back period can be less than two years, although these are the exception rather than a general rule.
4. Industrial robots are expensive to service and maintain. Again debatable, but obviously will vary with different types of robots and different applications. Quoting from one manufacturer, their robots require only three hours down-time for maintenance every three thousand working hours.
5. With continued rapid development, an industrial robot bought now could be out of date in two or three years time. Probably true, but if it does the job *now*, it will probably have paid back its cost in that time. Also there is no reason why it should not continue to be used. It does not necessarily have to be replaced by a more sophisticated robot as it can be reprogrammed repeatedly (within the limits of its original design).

CHAPTER 7

Power versus *Weight*

The heavier anything is the more power is needed to move it, raise it or perform any other mechanical function that is opposed by gravity and friction. This is a particularly important consideration in the design of mobile robots. Here we can start with an interesting comparison between two actual types.

The Westinghouse *Elektro* stands 7 ft. tall with a body constructed from steel tubes and frames covered with a sheet aluminium skin. *Elektro* weighs 260 pounds, and the total power provided by the numerous electric motors which operate his various functions is almost exactly 1 horsepower.

Peter Holland's *Mr. Robotham the Great*, demonstrated at the 1980 Model Engine Exhibition, is 6 ft. 3 in. tall, constructed largely from slabs of expanded polystyrene (foam plastic) covered with metallic paper. He has the same sort of metal man appearance, but weighs only $7\frac{1}{2}$ pounds. His power demands are very modest – a mere 1/40th horsepower as an approximate figure.

Thus direct comparison is very interesting. *Mr. Robotham the Great* weighs around 1/40th that of *Elektro* – and needs about1/40th the power. In other words the power to weight ratio of these two extremes of constructional weights is about the same – approximately 1/250th horsepower *per pound* of robot weight.

There are, of course, differences in the number and type of functions and movements these two robots are capable of performing; also in the manner in which they are performed. Heavyweight *Elektro*'s movements are slow and ponderous. Lightweight *Mr. Robotham* is quick and sprightly in his movements – simply because smaller masses can be accelerated more rapidly with less power. Where *Mr. Robotham* does lose out, however, is that his very light weight makes him susceptible to being blown over in a wind, and topple over easily on undulating surfaces or in collisions with other

1. Lunokhod 1, a Russian robot crawler-vehicle, landed on the Moon in 1970, carried there by Luna 16. It continued to move around the lunar surface for a year, sending back useful information.

2. The Mars Rover (*below*) — shown here in model form — is destined to be landed on Mars in the future. Nuclear powered, it will employ an 'intelligent' microcomputer brain to manoeuvre around obstacles and avoid hazards.

3. (*left*) Peter Holland's 'Mr Robotham the Great' is a 6ft. 3in. tall walking, talking robot who can also shake hands, bend knees to sit, swivel its body to face people. All functions are operated by radio control. Built of very light materials, it weighs only $6\frac{1}{2}$ pounds.

4. (*below*) Battelle's robot Mouse, featured on the front cover, is described in detail on page 7.

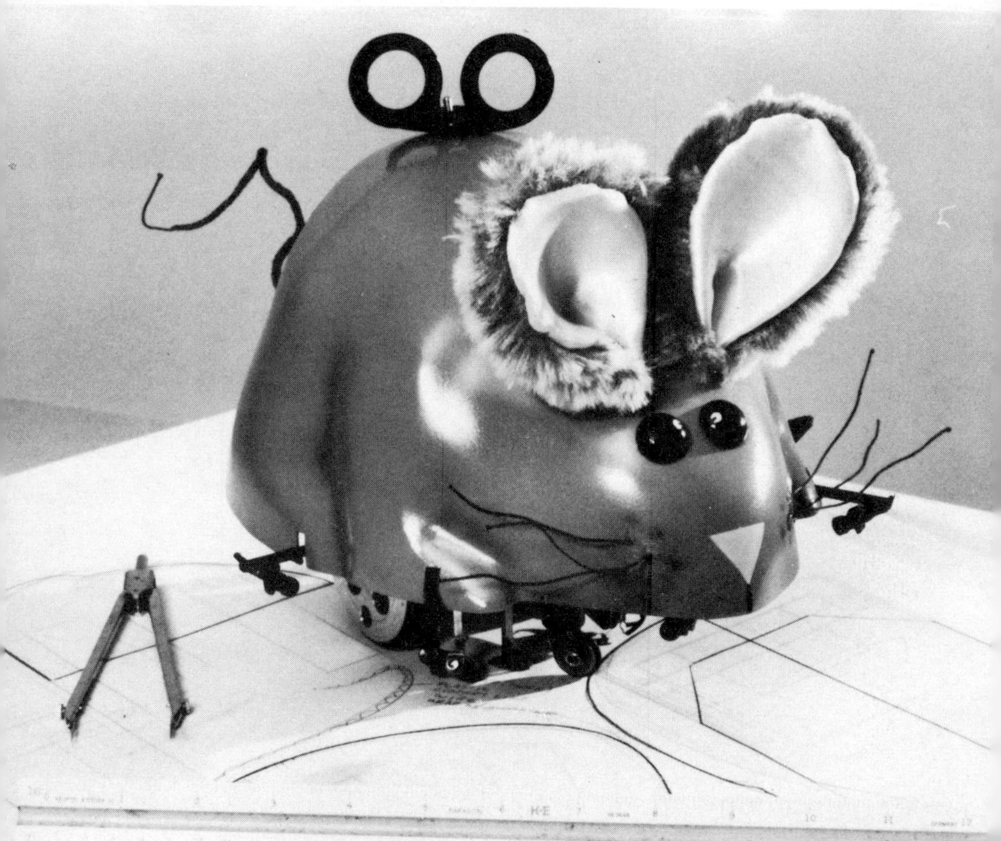

5. (*above*) Daros PT300V is an industrial robot from Dainichi-Sykes, shown here working as a lathe operator. It can work from a variety of positions around the machine.

6. (*below*) Daros PT800 industrial robot is a more powerful type capable of lifting heavy components onto a machine and removing them when the operation is completed.

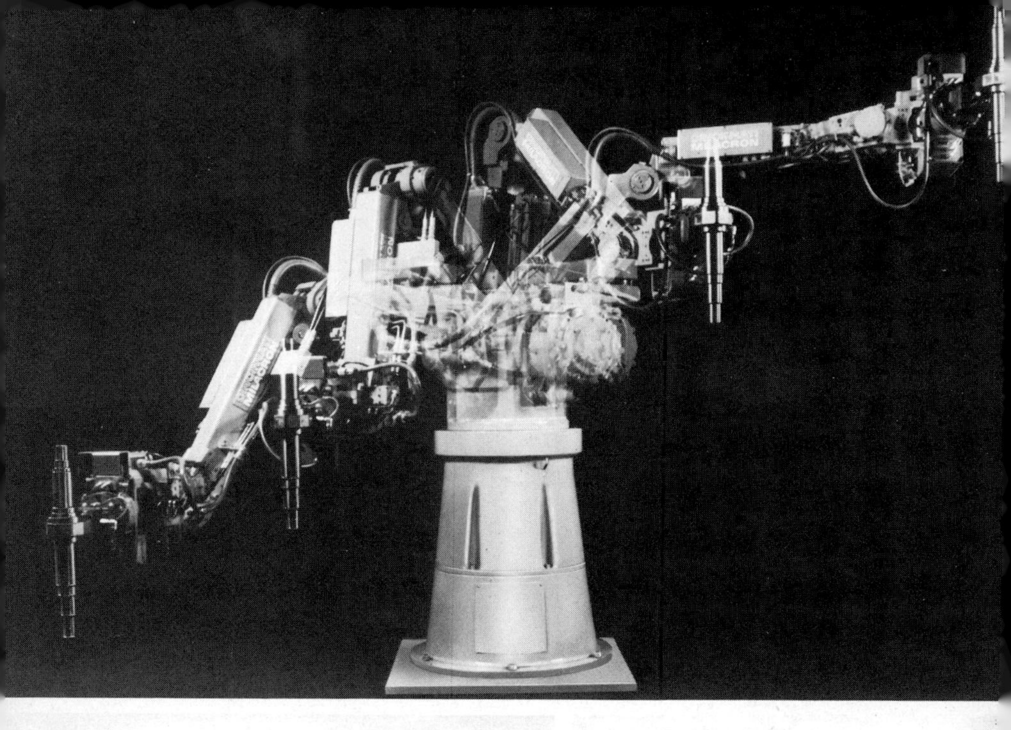

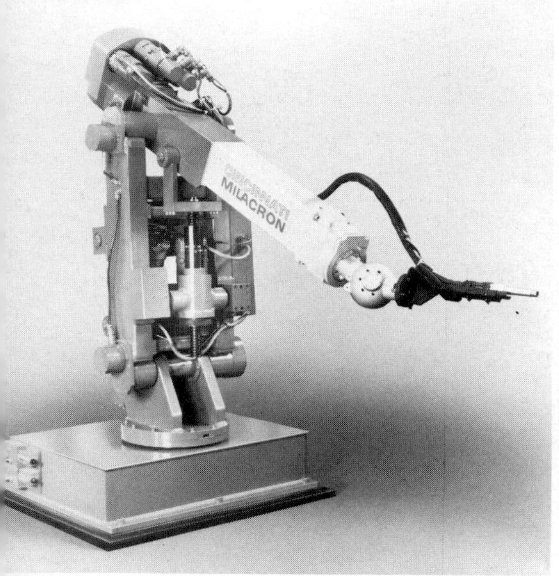

7. (*above*) The illustration is a multiple-exposure photo of a Cincinnati Milacron industrial robot showing the various degrees of freedom of the arm.

8. Another version of the Milacron is shown *below left* with the grip holding a spray gun.

9. (*below right*) JOBOT 10 is a general-purpose industrial robot adaptable to a wide range of handling or 'pick-and-place' duties.

10. Experimental computer-vision system being developed for the next generation of 'intelligent' robots. A TV camera at the rear views the moving conveyor belt beneath it, so that the computer 'sees' a digitized picture of the parts as they pass (inset photo *top right*). The computer then identifies individual parts from its pre-programmed memory and commands the robot arm to handle them accordingly.

11. (*above*) A 'seeing' robot's computer brain which filters out all but the essential information needed by the computer — the location of a part on a conveyor belt.

12. (*below*) The PUMA (Programmable Universal Machine for Assembly) is a small, relatively inexpensive industrial robot for handling and assembling small parts. It is shown here installing light bulbs of a car instrument panel.

13. (*above*) A robot welder at work. It has been 'taught' its work programme by a human operator and repeats it with complete accuracy each time.

14. (*below*) A line of industrial robots at work on a car assembly line. An important feature of programming here is that the welders must not get in the way of each other.

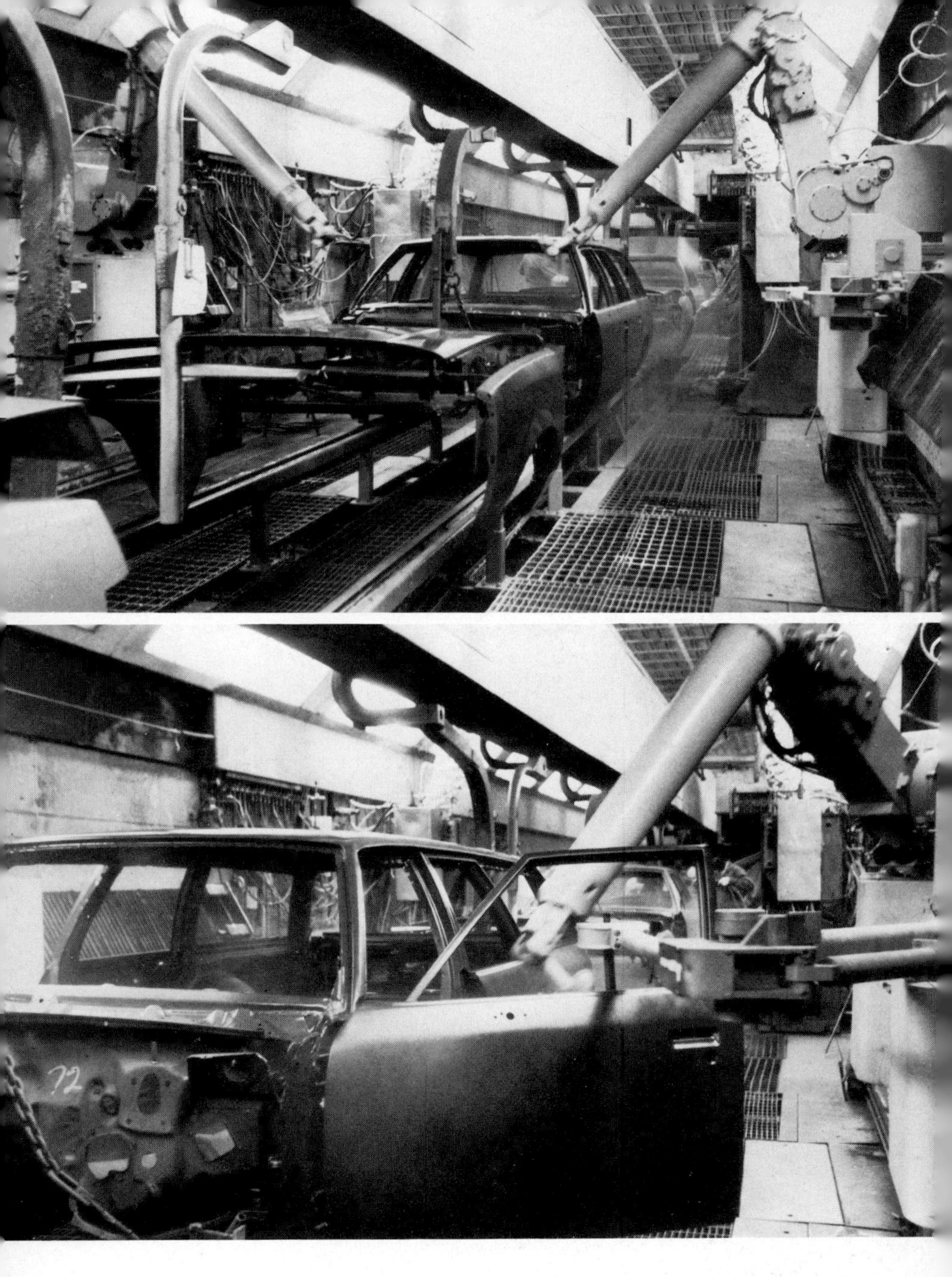

15. (*above*) A pair of NC (Numerically Controlled) robot painters are working together to spray the body and front assembly of a station wagon on an assembly line.

16. (*below*) Another robot joining in a painting session, this time to open and close car doors to allow interior painting.
Production lines of this type can operate with no human being present.

solution is to make the legs fully pivoted top and bottom in the form of a parallelogram linkage.

This will keep the feet level when striding. Further, if combined with another simple linkage system at the top to work as a motion divider leg movement can be made independent of the body, keeping the body upright and facing forwards all the time as the figure strides stiff-leggedly ahead – Fig. 6.7.

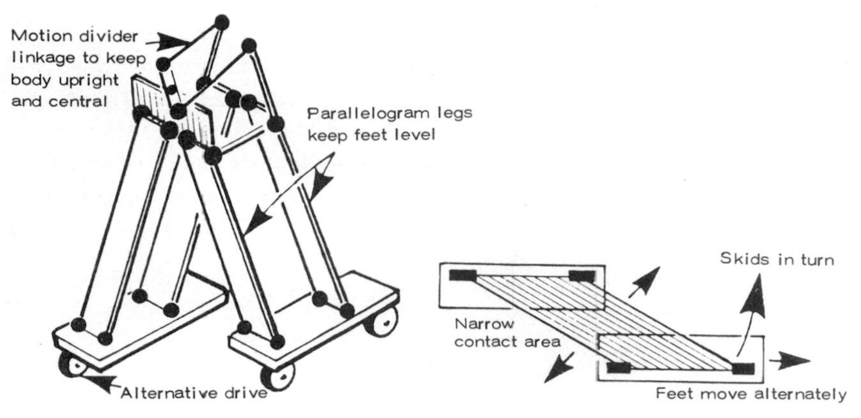

6.7 Peter Holland's ingenious parallelogram leg linkage with motion divider.

Even better is the 'walking plank' system of movement developed by Peter Holland in his hobby-type robots. The basic geometry of this movement is a rigidly linked system which progresses forwards in a series of arcs, at the same time maintaining a very stable base – Fig. 6.8. The wheels are enclosed with loose shoes, each guided by its own tracking wheel to move forward in a substantially straight line rather than an arc, i.e. the shoes hide the actual drive motion, giving the appearance of a reasonably natural 'walk'. Turns

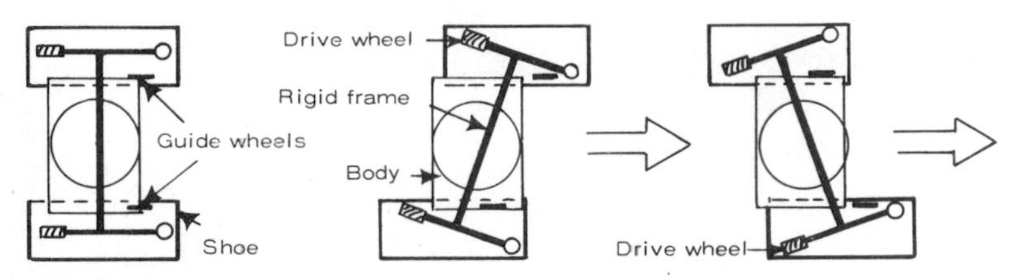

6.8 'Walking plank' system of movement – again devised by Peter Holland.

are made in the same way as the previous system — one foot motor driving with the other stationary, the figure then rotating about the stationary drive wheel. This movement is improved by having the support wheels (in the toe of each shoe) castor instead of having to skid when making turns.

It *is* also possible to make a biped (two-legged) robot walk up stairs. Peter Holland's *Tweeter* or robotic goose has a transverse track in the body with the legs connected some way in from the ends. The control payload and batteries traverse the track until the body becomes unbalanced and tilts, lifting the opposite side leg. The robot is then balanced on one foot, when the other foot can be rotated to move up on to the first step in front of it. Retraversing then transfers weight on to this foot again, when the other foot can be lifted again, and so on. These actions are performed by means of radio control.

CHAPTER 7

Power versus *Weight*

The heavier anything is the more power is needed to move it, raise it or perform any other mechanical function that is opposed by gravity and friction. This is a particularly important consideration in the design of mobile robots. Here we can start with an interesting comparison between two actual types.

The Westinghouse *Elektro* stands 7 ft. tall with a body constructed from steel tubes and frames covered with a sheet aluminium skin. *Elektro* weighs 260 pounds, and the total power provided by the numerous electric motors which operate his various functions is almost exactly 1 horsepower.

Peter Holland's *Mr. Robotham the Great*, demonstrated at the 1980 Model Engine Exhibition, is 6 ft. 3 in. tall, constructed largely from slabs of expanded polystyrene (foam plastic) covered with metallic paper. He has the same sort of metal man appearance, but weighs only 7½ pounds. His power demands are very modest — a mere 1/40th horsepower as an approximate figure.

Thus direct comparison is very interesting. *Mr. Robotham the Great* weighs around 1/40th that of *Elektro* — and needs about1/40th the power. In other words the power to weight ratio of these two extremes of constructional weights is about the same — approximately 1/250th horsepower *per pound* of robot weight.

There are, of course, differences in the number and type of functions and movements these two robots are capable of performing; also in the manner in which they are performed. Heavyweight *Elektro*'s movements are slow and ponderous. Lightweight *Mr. Robotham* is quick and sprightly in his movements — simply because smaller masses can be accelerated more rapidly with less power. Where *Mr. Robotham* does lose out, however, is that his very light weight makes him susceptible to being blown over in a wind, and topple over easily on undulating surfaces or in collisions with other objects. In other words, although light weight reduces power demands, it does have some penalties.

Battery Power

With mobile robots of this type, and the domestic robot, it is virtually essential that the complete power system be self-contained. Also the obvious choice of power unit is the electric motor, which means the robot must accommodate a matching set of batteries, recharged by plugging in to a suitable source of electricity when required. Powering internal electric motors from a trailing lead plugged into a mains supply would simplify and lighten the system, but at the expense of reduced mobility. A practical domestic robot with a trailing electric lead, for example, could all too easily get tangled up in its manoeuvring from place to place.

Accepting internal battery power as virtually essential, the effect of robot weight as affecting power requirements is again significant. The lower the power requirement the easier it becomes to *reduce* weight by reducing battery size and weight. In other words, if you need more power, you will also need proportionately bulkier and heavier batteries.

Here, too, a look at different battery types can be instructive. For plenty of 'battery power' a lead-acid battery is the normal choice. This is the type used in electric lawnmowers and most smaller electric-powered vehicles. Where only low electric power is required, other types of batteries — and particularly the nickel-cadmium battery — become more attractive both on an *energy density* basis and *energy/volume ratio*.

The energy density of a battery is defined in terms of watt-hours of electrical energy it can produce per unit weight. Thus where a minimum weight battery is required, the logical choice is a battery with the highest energy density.

Typically a lead-acid battery may have an energy density at worst of about 2 watt-hours per pound weight; and at best 20 watt-hours per pound weight, depending on construction. For example, a typical motor cycle battery could have an energy density figure between 8 and 15 watt-hours per pound. These figures could be bettered, but not by very much, by batteries specially designed and constructed for automotive traction.

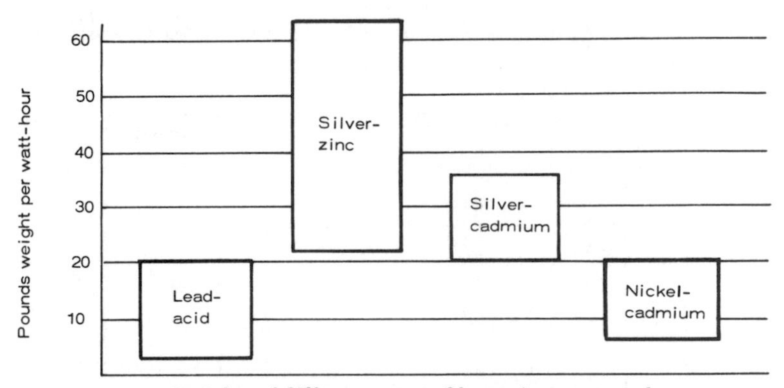

7.1 Weights of different types of batteries compared.

Energy density figures for some other battery types (*see also* Fig. 7.1) are:

Silver-zinc — 23–65 watt-hours per pound, or say 53 watt-hours per pound as typical.

Silver-cadmium — 20–35 watt-hours per pound, or say 33 watt-hours per pound as typical.

Nickel-cadmium — 6–20 watt-hours per pound, or say 13 watt-hours per pound as typical for standard types; and 15 watt-hours per pound for sintered plate types.

On this basis the nickel-cadmium battery does not show any particular advantage over a lead-acid battery. Also it is more limited in size range available, and thus capacity. The silver-zinc battery on the other hand, and to a lesser extent the nickel-cadmium battery, show definite superiority in energy density. Their very much higher cost, however, precludes their use for all but highly specialized applications. The same is true of more recent battery systems, still largely in the development stage, which can also show a rapidly superior energy density performance. For general application, therefore, the choice virtually remains between the lead-acid accumulator and the (rechargeable) nickel-cadmium battery.

The *energy/volume ratio* of a battery is a guide to the minimum size of battery for a given application. If there is a need to choose a battery of smallest possible physical size, the one with the highest energy/volume ratio or *watt-hours per cubic inch* becomes the preferred choice. Here are some comparable figures (*see also* Fig. 7.2):

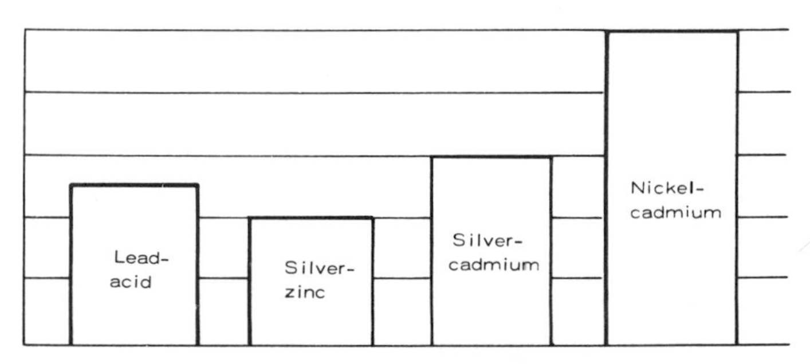

7.2 Sizes of different types of batteries compared.

Lead-acid batteries — 0.3–2.0 watt-hours per cubic inch, depending on type. Not a very useful guideline because of the wide range. Even with motor cycle batteries a likely range is 0.3–1.1 watt-hours per cubic inch; and for automotive traction batteries, 0.5–2.0 watt-hours per cubic inch.

Nickel-cadmium batteries — 0.2–1.4 watt-hours per cubic inch, but typically of the order of 1.0 watt-hours per cubic inch in modern designs.

Silver-zinc batteries — 0.7–3.5 watt-hours per cubic inch, with 2.5 watt-hours per cubic inch typical.

Silver-cadmium batteries — 1.6–2.2 watt-hours per cubic inch, with 1.7 watt-hours per cubic inch being fairly typical.

Watts into Horsepower

Despite the attempts by various Standards authorities to make watts the universal unit for power, the term *watts* is readily understood to mean electrical power; and *horsepower* the power output of 'mechanical' systems and energies. The electric motor is both a 'mechanical' and or 'electrical' system, so its output power may equally well be described in terms of watts *or* horsepower.

The relationship between watts and horsepower is:

$$1 \text{ horsepower} = 746 \text{ watts}$$
$$\text{or } 1 \text{ kilowatt} = 1.3 \text{ horsepower}$$

In the case of electric motors, power output quoted in watts can be misleading unless it is definitely understood that it refers to *output* power. Electric motor *input* power is also specified in watts (or kilowatts), watts being equal to input voltage × amps current drawn by the motor. Output power will always be less than input power. The efficiency of electric motors can range from 95 per cent or so for large motors, down to 50 per cent or less for small model-size types. Thus whilst it is easy to calculate the *input* power of an electric motor as volts × amps, this does not provide very accurate means of estimating *output* power unless the *efficiency* of the motor is also known. Also efficiency can vary with operating speed (and at the same time current drain will also vary with speed in the case of battery powered DC motors).

Thus a 12-volt motor drawing, say, a normal current of 5 amps is a $12 \times 5 = 60$ watts *input* motor. Its actual output power at different efficiencies could vary considerably, for example:

efficiency	output power in watts	horsepower
90%	54	0.072
80%	48	0.064
70%	42	0.056
60%	36	0.048
50%	30	0.040
40%	24	0.032

Apart from illustrating the effect of motor efficiency on power output, these figures also emphasize the large amount of electrical power *input* necessary to produce high power from electric motors. At 60 per cent efficiency, for example, an input of 360 watts would be necessary to produce a $\frac{1}{2}$ horsepower output. Considering this motor powered by a 24-volt battery, this would result in a current drain of 15 amps.

Battery Capacity and Duration

The *capacity* of a battery is expressed in terms of ampere-hours (abbreviated Ah), and is independent of voltage. Given the capacity, and knowing the current drain, then the theoretical time the battery will last before requiring recharge is simply capacity divided by current drain. Thus if a battery with a capacity of 7.5 Ah was used in the above example, its maximum duration of working would be 30 minutes. In practice, and particularly with high current drains, it would be less — say 60–70 per cent of the theoretical capacity or about 20 minutes.

The same sort of calculation can be applied to all types of battery, regardless of size. The exception is carbon-zinc dry batteries (the most common form of non-rechargeable battery). With this type it is not possible to specify a capacity figure as this can vary widely with current drain and frequency of use. But non-rechargeable batteries would not be seriously considered for robot power in any case (outside toys) because of their need for frequent replacement. Although higher in initial cost, rechargeable batteries are much more economic — and more reliable in the long run.

Power Requirements

It is virtually impossible to calculate power requirements to make a robot mobile from first principles. With any driven wheel system, for example, theoretically there is no friction between a wheel and a level surface over which it runs (unless the wheel slips). The only realistic quantity which can be calculated on this basis is acceleration performance:

$$\text{acceleration} = \frac{\text{final velocity} - \text{initial velocity}}{\text{time}}$$

or

$$A = \frac{V - V_0}{t}$$

If there is no initial velocity, i.e. acceleration is from a standstill:

$$A = \frac{V}{t}$$

Time is also related to distance covered:

$$\text{time} = \frac{\text{distance}}{\text{average velocity}}$$

or

$$t = \frac{D}{V/2}$$

What are realistic design figures for final velocity (V) and acceleration (t) for a mobile robot? If it is much faster than a man's walking speed (say 4 mph) it could be hazardous. If appreciably slower, the robot would appear sluggish. A possible design figure is thus around 4–5 mph top speed, or say 5 ft/sec.

The size of the robot also comes into this thinking, which applies to a 'man size' robot. A larger robot could be even more of a hazard travelling faster, so 6 ft/sec. is still a good design figure. A smaller robot will have increased stability problems travelling at the same speed, and would be more 'realistic' moving about at a lower speed. On the other hand, smaller creatures move faster than man, so again our 5 ft/sec. speed figure seems about right.

Man can accelerate very rapidly from a standstill to top walking speed. Too rapid an acceleration with a wheel driven robot could again be dangerous (especially with a heavy robot), at the same time presenting control difficulties and stability problems (i.e. make it liable to tip over backwards). Here again we can only 'guesstimate' and a suggested figure is to reach maximum velocity over a distance of 6 feet. Then:

$$\text{Time to accelerate to top speed} = \frac{6}{5/2} = 2.4 \text{ seconds}$$

$$\text{Acceleration} = \frac{5}{2.4} = 2.08 \text{ ft/sec or say } 2 \text{ ft/sec}^2$$

The corresponding *force* required to produce this acceleration can be calculated from the basic formula:

$$\begin{aligned} \text{force} &= \text{mass} \times \text{acceleration} \\ &= \frac{\text{weight}}{32.2} \times \text{acceleration} \end{aligned}$$

An acceleration figure of 2 ft/sec^2 has already been estimated. Suppose we apply this to a robot weight of 50 pounds:

$$\begin{aligned} \text{Force} &= \frac{50}{32.2} \times 2 \\ &= 3.1 \text{ ft-lb.} \end{aligned}$$

Power is the product of force and velocity. One horsepower is equal to 550 ft-lb per second, giving the equivalent equation:

$$\text{Horsepower} = \frac{\text{force} \times \text{average velocity}}{550}$$

Or in this case

$$\text{Horsepower} = \frac{3.1 \times 5/2}{550}$$
$$= 0.0147$$

This represents a very modest power requirement. In practice higher power would be required to overcome friction, which in turn would largely depend on the surface over which the robot is intended to travel. There is no real way of estimating this, but there is a trick we can employ to calculate power required under simulated 'extra resistance' conditions. That is to determine the force, and hence the power required to drive the robot up an incline.

This additional force is given by:

F = weight × sine of incline angle.

Here again we have to 'guesstimate' a realistic value for the incline angle, say:

(i) 5–6 degrees for a robot designed to operate over smooth surfaces, when we can adopt a figure of 0.1 for the sine of the incline angle.

(ii) About 12 degrees for a robot designed to operate over rougher surfaces and even negotiate inclines up to about 5–6 degrees. Here we can adopt a figure of 0.2 for the sine of the incline angle.

Still considering our 50 pound weight robot, the additional forces present under the above conditions are:

(i) 50 × 0.1 = 5 lb
(ii) 50 × 0.2 = 10 lb

Add these to the previously calculated acceleration force (3.1 ft-lb) and recalculate horsepower requirements:

(i)
$$\text{HP} = \frac{(5 + 3.1) \times 5/2}{550}$$
$$= 0.037$$

(ii)
$$\text{HP} = \frac{(10 + 3.1) \times 5/2}{550}$$
$$= 0.06$$

Basic calculations like these can be very useful to estimate likely power requirements – substituting any required figures for acceleration, top speed and solid weight – or definite ability to climb a particular incline. Power requirements will increase considerably with larger weights and high accelerations and top speeds. Equally such calculations can be worked the other way round to estimate likely performance starting with a given size and power of electric motor.

Alternative Calculations

The foregoing calculations have been based on first principle formulas. There is an alternative method of approach, basing calculations on *energy* requirements:

Acceleration up to a given velocity from a standstill represents a change in kinetic energy equal to:

$$KE = \frac{W}{2g} \times V^2$$

Apply this to the original design figures:

$$W = 50 \text{ lb}$$
$$V = 5 \text{ ft/sec.}$$
$$\text{Hence } KE = \frac{50}{2 \times 32.2} \times (5)^2$$
$$= 19.4 \text{ ft-lb.}$$

The relationship between kinetic energy and horsepower is:

$$\text{Horsepower} = \frac{KE}{\text{time} \times 550}$$

The time, from previous calculation, is 2.4 seconds

$$\text{Hence Horsepower} = \frac{19.4}{2.4 \times 550}$$
$$= 0.0147$$

(i.e. exactly the same value derived from the previous calculations).

If climbing a gradient is taken into consideration, an additional change in energy is involved – the *potential energy* (PE) resulting from the height gained in climbing a specific distance up the slope.

This is given by:

$$PE = W \times \text{distance} \times \text{sine of incline angle}$$

The distance has been previously determined as 6 feet.

Thus for the nominal 5–6 degree incline (sine = 0.1)

$$\begin{aligned} PE &= 50 \times 6 \times 0.1 \\ &= 30 \text{ ft-lb} \end{aligned}$$

Add this to the charge in kinetic energy (KE) previously calculated:

$$\begin{aligned} \text{total charge in energy} &= 19.4 + 30 \\ &= 49.4 \text{ ft-lb.} \\ \text{when Horsepower} &= \frac{49.4}{2.4 \times 550} \\ &= 0.037 \end{aligned}$$

Or for the nominal 12 degree incline (sine = 0.2)

$$\begin{aligned} PE &= 50 \times 6 \times 0.2 \\ &= 60 \text{ ft-lb.} \\ \text{where total charge in energy} &= 19.4 + 60 \\ &= 79.4 \\ \text{when Horsepower} &= \frac{79.4}{2.4 \times 550} \\ &= 0.06 \end{aligned}$$

These results are exactly the same as those determined by the original calculations, so it does not matter which method is used. It is a good idea, in fact, to use both methods, when one set of calculations will serve as a check on the other.

CHAPTER 8

Robot Senses

Desirable features of any more advanced type of robot are that it should be able to 'see' and 'feel'. To what extent it needs to 'see' and 'feel' depends entirely on the duties it is designed to perform. Thus different types of robots need different degrees of 'seeing' and 'feeling', and may or may not need both senses.

The third basic sense is an ability to communicate. This does not necessarily mean via speech (although a robot can easily be designed to respond to spoken commands, and there are also robots that can talk back). It means the adoption of a suitable language via which robots can be instructed and programmed, normally via an electronic brain in the robot. Basically, therefore, this means computer-type language, but in as simplified a form as possible. It is only the programmable robots that really justify the description of being true robots, and these should also have the ability to accept instructions to change programmes.

'EYES' FOR ROBOTS

The simplest and most obvious choice for a 'seeing eye' for a robot is a photo-electric cell or phototransistor. Both are small electronic devices which if used as part of an electric circuit, the value of the current flowing through the circuit will vary with the amount of light following on the photocell — Fig. 8.1.

The simplest mode of working such a system is to use a single

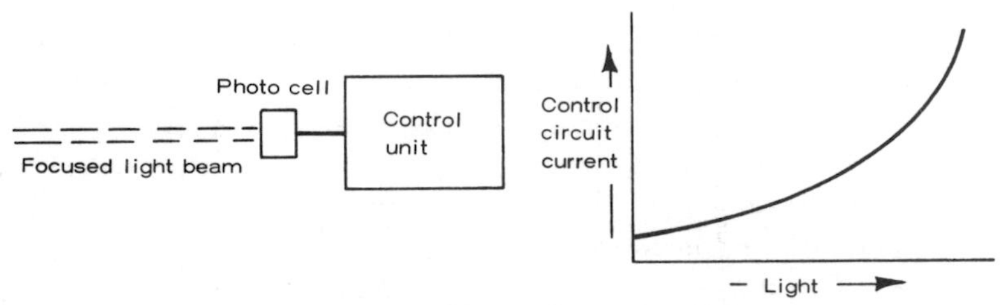

8.1 Reaction of photocell circuit to light.

(photocell) eye and a focused light beam. Assume first that the robot merely pivots about its base, this motion being produced by an electric motor. The objective is to make the robot turn to face the source of the light beam, when current flowing through the photocell circuit will be at maximum. This basic circuit also includes other components controlling switch on and off and direction of rotation of the motor. With *maximum* current flow, the motor is switched *off* (because the eye is now directly facing the light). In any other position of the eye, the control circuit drives the motor in such a direction that circuit current *increases*, i.e. the eye will always seek the light position and stop — Fig. 8.2. Equally, if the robot is a mobile type, the same principle can be used to make the robot home in on the source of light, following an undulating path. In exactly the same way a robot can be made to point towards or home in on a *moving* source of light; also since 'low' current always means 'turn in the direction which increases current' as far as the control system is concerned, such a single eye has 360 degrees 'vision'.

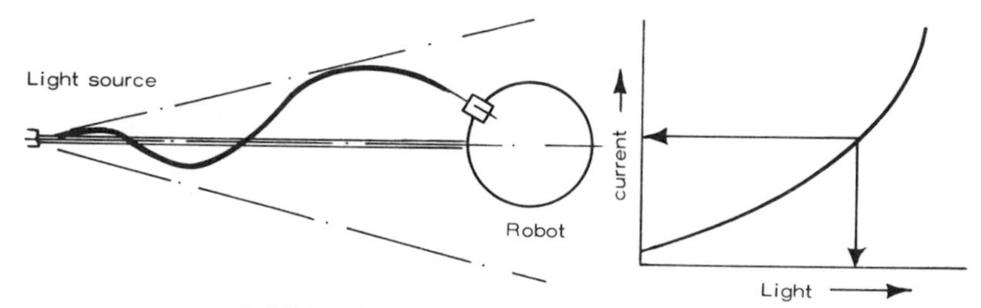

8.2 'Homing' system using single photocell.

On the basis that two eyes can be better than one, there are some interesting alternatives worth considering. One is to use a very narrow light beam with two eyes which come outside this beam when directly facing the light source — Fig. 8.3. If the robot is turned away from the beam until one eye enters the beam, this immediately gives maximum signal current in that circuit which is used to drive the motor to turn the robot in the opposite direction. In this way the robot continually corrects any deviation from a position pointing towards the light source. It has the advantage of needing a simple control circuit, i.e. the need to drive the motor only when *maximum* current is flowing in that circuit; and with two separate circuits (as for each eye), switching the motor in opposite directions is simple.

Such a system, however, only has limited *forward* vision, not 360-degree vision.

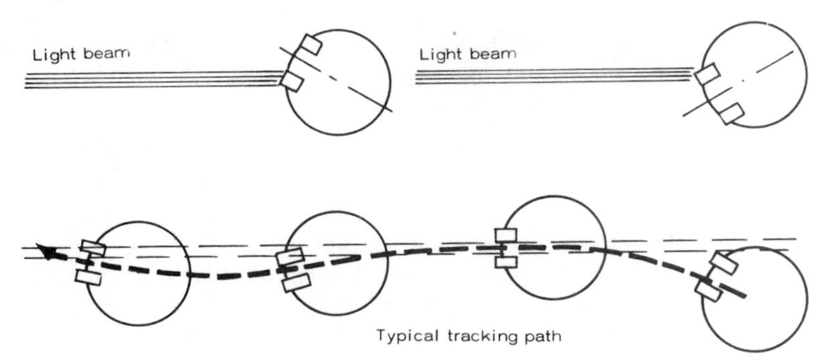

8.3 Two eyes (photocells) can be better than one.

Another two-eyed system is shown in Fig. 8.4. In this case the eyes (photocells) are mounted on the back of a prism, one each side. Facing directly towards the light source, light is reflected around the inside of the prism and back out again. None reaches either eye. If light comes from the left, however, it will enter the prism and fall on the right eye. This imitates a control signal telling the robot to turn to the left. Similarly light coming from the right enters the left eye, producing a signal for a turn to the right. Command signals are again a maximum when initiated, and separate signals for left or right turn. Again, however, such a system has limited 'forward' vision.

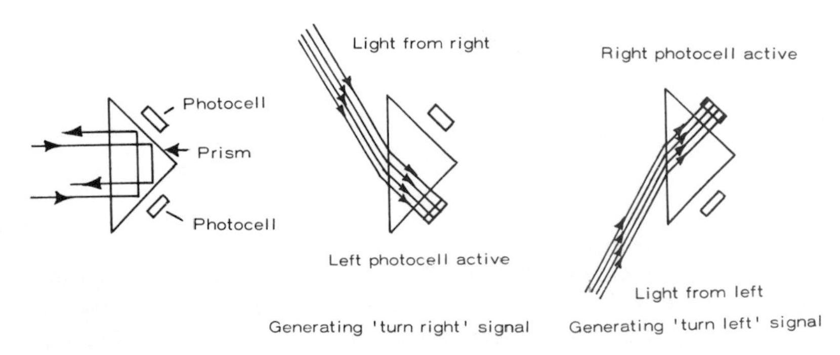

8.4 Two-eyed prism system.

A limitation of all photo-electric eyes is that the intensity of the light reaching the cell, and thus the circuit current, decreases with the square of the distance of eye to source. In other words, the further away the robot is, the weaker the command signals will be. It is essentially a short-range eye, and dependent on the light source being focused into a relatively narrow beam.

One way in which this can be overcome is to use high intensity pulsed light signals. To avoid any danger to human eyes looking at such light, the robot's eyes would then need to be mounted low down under its skirt, with the light source also at floor level.

Another alternative in the case of a mobile robot is to mount both light source and eye on the robot, directed vertically downwards so that the light beam is bounced back off the floor in the direction of the eye — Fig. 8.5. The path the robot has to follow is then laid out by a strip of metallic tape or reflective white tape. The eye then seeks out the position of the tape (maximum signal current) and follows it.

This same principle is widely used on industrial robots for positioning, i.e. the robot arm controlled by signals from a photocell eye seeks out a 'light' or 'dark' area to locate on. In this manner the eye can detect the rim of a component from its body, say, or fix on the right point to grip it or work on it.

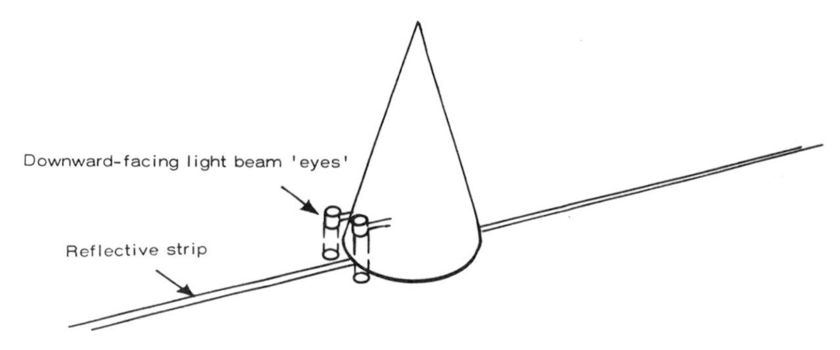

8.5 Downward facing 'eyes' for following a path laid out on the floor.

The Video Camera Eye

At first thought this would seem the complete answer to providing a robot with sight: a miniature video camera which provides a complete picture of what it is facing. Unfortunately a video picture is flat (two-dimensional) and changes with position. An obstruction in the form of a rectangular panel, for example could appear on a video screen as a stick, viewed edge-on, changing into an upright rectangle, square and then a horizontal rectangle viewed from different aspects in a 90 degree arc (and repeated three times over if the robot circled the object).

This is only a simple situation. In realistic applications, many objects are likely to be in the picture overlapping each other in different points of view. Moving objects mixed with stationary objects would generate even more picture variations.

Identification of three-dimensional objects via flat screen TV would, therefore, need complex electronic brain power to identify fully — far greater than can be accommodated economically, or even practically. Nevertheless video camera eyes are used on certain types of industrial robots working in hazardous areas, but only to let a human operator watch what is going on from a safe, remote distance.

Where Robot Eyes can be Superior

Robot eyes can, however, have a superior performance to human eyes in certain circumstances. They need not be simple light sensors. They can be designed to detect infra-red radiation and thus 'see' in the dark, and even detect or identify heat sources, materials, radiation, etc; even accurately measure the distance objects are away from them.

A relatively simple electronic eye, for example, could readily distinguish between metallic and non-metallic objects (which human eyes can also normally do by appearance); and also distinguish between magnetic and non-magnetic metals, e.g. between aluminium and steel even if both were painted over (which the human eye could not do).

Carry this to an extreme, and there is already in existence a robot eye which can scan, and virtually immediately, identify *any* metal simply by looking at it. The robot eye in this case is basically a spectrometer, connected to an electronic brain which immediately identifies the metal by the number and position of the lines seen by the analytical eye. There are other types of robot eyes where the eye is basically a scientific instrument connected to a microprocessor brain for immediate read-out of observations.

Any further extension of the 'seeing' ability which can be

incorporated in robots properly comes under the heading of *sensors*, which is the subject for a separate chapter.

Sensors also include the various methods and systems whereby robots can be given a sense of feel.

SPEAKING AND LISTENING ROBOTS

A built-in capability for a robot to respond to spoken commands, or itself speak in a human voice, are both novelty features which are attractive to include in demonstration and hobby type robots; and will probably be regarded as essential in domestic robots when and if they eventually appear as practical, readily available productions. Communication by speech is unnecessarily limited. It is essentially a foreign language as far as effective functioning of first- or even second-generation robots is concerned, although there are a few exceptions.

These include artificial noise or *speech processors*, primarily developed as teaching aids. A talking mini-computer is a device of this nature — a robot brain with a speech capability. There have been further developments on this basis, dressing up the robot brain in a stationary robot figure to give it a more human association. It has, in fact, been found that slow-to-learn children can respond to these better than to a human teacher.

The qualification to be described as a true speaking robot is a synthesized electronic noise generator together with a pre-programmed vocabulary of word and sentences which can be called up in sequence, normally on a question-and-answer basis. Such a programme incorporates pauses between answers to ask a question, similar to communicating with a computer in BASIC language. Anything less is not true robotic speech, but there are other obvious — and simpler — methods of making a robot 'talk', e.g. via a tape recorder fitted in the robot; or reproducing speech from an operator via a radio link with a receiver and loudspeaker in the robot. Both are systems only deserving consideration in demonstration, display or talking type robots. The tape recorder system is not robotics, even though the speech is programmed in the sense that it is pre-recorded. The radio transmission system is merely a form of ventriloquism.

Anything more than the pre-programmed BASIC computer type speech capability with synthesized noise is likely to have to await third-generation rather than second-generation robots (and second-generation robots are only just beginning to appear). The possibilities are somewhat awesome. A talking computer that thinks out its own replies!

Listening Robots

By contrast the robot that responds to spoken commands merely follows well-established techniques, employing a microphone to convert received sound into command signals. This circuit can incorporate audio filters, so that it responds only to a particular type of sound (i.e. particular frequency content), such as a hum or other noise pattern. Equally it could be made to respond to a whistle, or musical note, etc., or be made even more selective in the frequency content and thus the type of noise.

The simplest control system then works by *sequence* using one syllable words, i.e. one spoken word produces response by the first action in the programmed sequence; the next word the second action; and so on. The sequence may be *repetitive, cancellable* or *selective*. In the former case the robot has to work through the complete programme before it can restart the same programme. This is clumsy unless the programme is restricted to a few words and actions only. For example, consider a simple four word programme:

(i) First word — e.g. 'Go!' — makes the robot move forwards.
(ii) Second word — e.g. 'Left!' — makes the robot turn left.
(iii) Third word — e.g. 'Right!' — makes the robot turn right.
(iv) Fourth word — e.g. 'Stop!' — makes the robot stop.

Suppose after commanding 'Go!' followed by 'Left!' there is an immediate need to make the robot turn left again. To get to this sequence the remaining sequence, plus the start signal, and left turn signal has to be worked through. This would require the spoken command:

'Skip' (iii) — 'Skip' (iv) — 'Skip' (i) —
and finally 'Left!'

Command programmes like this are easy to write, and quite easy to translate in terms of robot control circuitry using a rotary switch which achieves a step at a time when each pulse (word) signal is received via a microphone. For example, each word 'heard' by the microphone generates a positive pulse in the control circuit which is used to step the rotary switch to the next position in the sequences – Fig. 8.6.

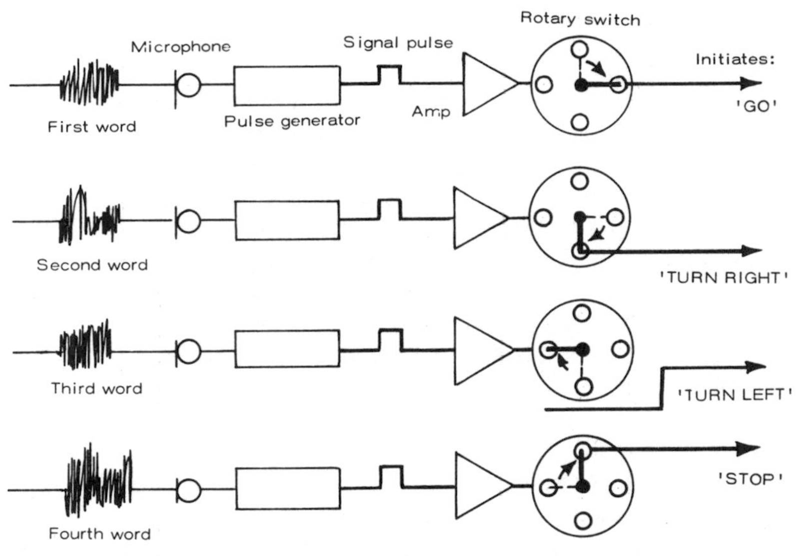

8.6 Basis of sequence system for commanding a robot by spoken words.

In practice, it is a tricky control system to use effectively – the greater the number of sequences employed the more readily the sequence is lost. A *cancellable* sequence control system is little better. Here, a separate command said – say a whistle – is used to trigger a separate circuit, returning the sequence to its initial or 'safe' position, e.g. stop. Thus, if a sequence is lost, the operator can give the 'Cancel!' signal and start all over again.

A *selective* sequence system does not have this basic limitation of losing sequence. Command signals in this case result in the sequence switch in the robot moving *directly* to its appropriate place. This can still be done on a pulse basis. For example, taking the original four sequence programme:

(i) 'Go' – signal one pulse – e.g. speak 'Go!'
(ii) 'Turn left' – signal two pulses – e.g. speak 'Turn left'
(iii) 'Turn right' – signal three pulses – e.g. speak 'Turn to Right!'

(iv) 'Stop' – signal four pulses – e.g. speak 'Now Stop and Think!'

Such a programme will not work properly unless it has a method of *terminating* a command and is also *cancellable*. Both features can be incorporated in the 'Stop' command. Thus the robot is commanded to 'Stop' when it has moved far enough, or turned enough, and at the same time the sequence switch is automatically returned to its initial 'awaiting signal' position ready to receive the next command. Without this cancelling feature, sequence will be lost if one command is immediately followed by another.

Basically controlling a robot by speech commands is a gimmick, used only in demonstration and hobby type robots for effect. Radio control, for example, is far more positive, and can also readily give *proportional* movement responses.

CHAPTER 9

Sensors

A sensor is defined as a device for the detection or measurement of a physical property to which it responds. As far as robotics is concerned this covers an ability to derive input signals relative to tracking, attitude, distance or proximity, grip and any other specific 'senses' that an individual robot is designed to have. In effect they provide data relative to a question posed by the robot's immediate position, such as 'Which way to travel?' or 'How near is that object?', and so on. Such signals are then translated by the robot's 'brain' into command signals to take what action may be necessary.

There is also another type of 'sensing' known as *feedback* used to measure and correct any position error. If the command is not executed properly, e.g. the robot does not react, or overruns, a commanded position, an error signal is produced which works to correct the position. When this is reached, the error signal falls to zero as no more correction is needed. Feedback, therefore, can be an essential feature of *positioning* controls, but may not be necessary on others.

Tracking Sensors

The photo-electric 'eye' is the popular concept of a tracking sensor applied to robots, as described in the previous chapter. It is capable of giving the robot a sense of sight, although this is distinctly limited in scope. It is essentially a short-sighted eye, but useful in many applications none the less.

Much simpler solutions are possible if a particular track is to be followed repetitively. For example, the robot can simply run on, or be suspended from, rails, needing no sensor at all. This is a solution adopted on many industrial robots that need to move from one place to another.

Another relatively simple alternative is to lay the 'track' in the form of a wire or strip of metal laid on, or buried just under the surface over which a mobile robot is to operate. An electric current is fed through this wire, generating a magnetic field around it. The robot is fitted with a sensory coil which seeks out and tracks the robot along the path of the wire on an 'error signal' basis — Fig. 9.1.

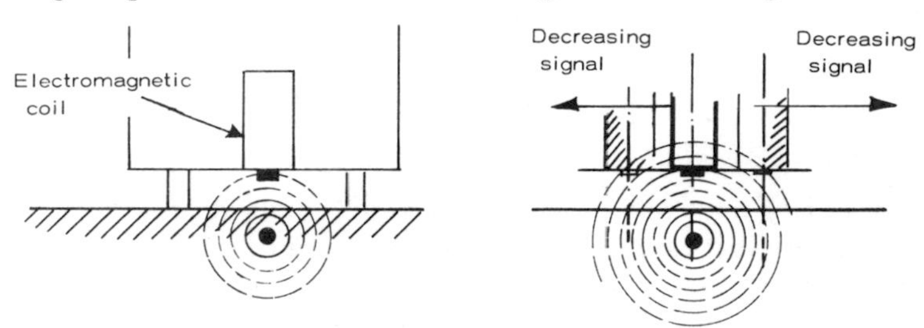

9.1 Buried wire and electromagnetic coil sensor — a system widely used with mobile industrial robots.

This system is particularly suitable for mobile industrial robots where the paths they are to travel can be clearly defined along clear-ways. The robots have right of way along such paths, which, for the benefit of humans working alongside robots, are marked out in paint on the floor. It is up to the human worker to keep clear of moving robots, for about the only concession the robot makes to human safety is to flash hazard lights immediately prior to, and when moving.

Thus the working robot can be far removed from a droid in concept. This is quite acceptable in an 'all-robot' factory where few, if any, human operators are needed on the shop floor. Collisions between robots themselves are eliminated by a properly programmed robot choreography. In a 'mixed' environment, with robots working alongside humans, robots would need the addition of a proximity sensor so that they would stop automatically on nearing an obstruction.

One or Two Sensory Coils?

In theory, at least, a single coil can provide such control signals on the basis that there will be maximum signal strength when the coil is directly over the wire. That is, any departure from this tracking will reduce the signal received, which is interpreted as an error. The control circuit receiving the signal from the sensory coil translates this into a correcting steering command until the sensory coil is generating maximum signal strength again.

In practice, such a tracking control can be quite vague because the position generating maximum signal is diffused rather than sharply defined. Two (or even three) sensory coils can provide much more precise tracking — Fig. 9.2.

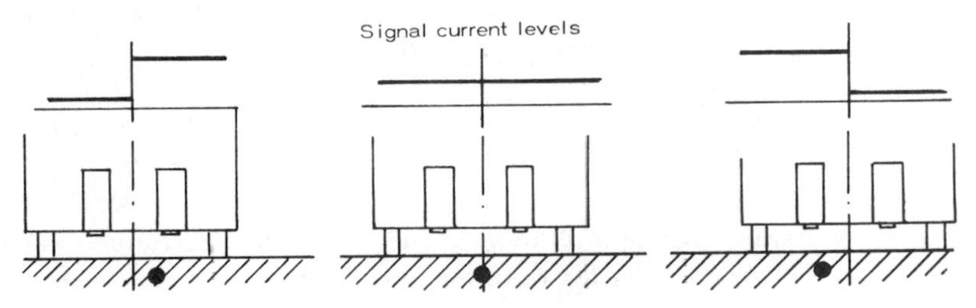

9.2 Electromagnetic sensing is improved using two sensing coils.

With two sensory coils each will generate a signal strength proportional to the distance from the wire. Thus if the robot is not centralized over the wire, the signal from one sensory coil will be greater than that from the other. A simple comparator circuit can measure this difference or error signal and apply this to command the steering to correct itself. Another advantage of using two sensory coils is that the more off track the robot is, the stronger will be the error signal.

Typically such a tracking system can operate quite effectively with only a low current required in the wire. AC is more effective than DC in this respect, and an AC circuit system with a frequency of 1000-5000 Hz would only need a current of about 150-200 milliamps flowing through the wire. With three sensory coils, one centrally located, the system can be even more effective, using less current still.

Proximity Sensors

Sensing of proximity or closeness to an object falls into two distinct categories — distant and close-up. These involve the use of quite different types of sensors. Also, for most practical applications, 'distant' sensors are more of a luxury than a specific requirement.

The most positive type of 'distant' sensor is one operating on the echo-sounder principle. This involves a transmission of a suitable signal which is reflected back off the distant object, picked up by a receiver, and the time taken for the signal to travel to the object and back is read out as a measure of the *distance* between the object and the sender.

Possible *types* of signals are *sound* in both the audio- and ultrasonic frequency ranges, *microwave radio* signals (radar), *light* and *infra red* radiation. Sound is the simplest to use since a single piezoelectric transducer can act both as a transmitter and receiver of signals. Microwave radio is complicated by the fact that it requires both a transmitting and receiving aerial, although these can be combined in one unit, and the circuitry involved is much more extensive and complex. Ordinary light beams are quite unsuitable, although limited distance measurement can be made with stroboscopic light (using the same principle employed in automatic electronic flash units for cameras). A *laser* beam is much more effective, with a distance-measuring capability similar to radar, but potentially dangerous as it would blind anyone accidentally looking into the laser beam, or even the reflected beam. It could be a practical system in an all-robot factory, but not for distance-sensing robots with humans in the vicinity.

'Close up' sensors are true proximity sensors in that they detect closeness to (an object), rather than actual distance from it — an essential feature in a robot arm, for example, which has to position itself before it can undertake a task. They are similar to *tactile* sensors, but with one important difference — they sense proximity without actually touching the object involved.

The photo-electric 'eye' can be used as a proximity sensor, working on a light/dark basis, i.e. it is made to sense a light or dark area on the object as a control point. The 'sight' of such an eye does not have to be direct. It can be extended via parallel bundles of fibre optic materials so that the 'eye' is positioned at the most convenient close-up point — *see* Fig.9.3.

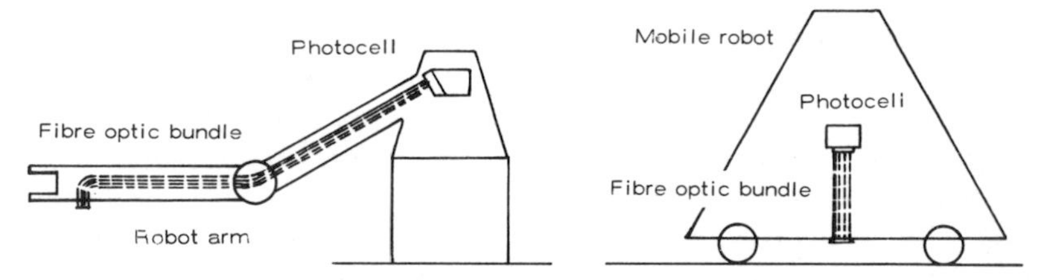

9.3 Using fibre optic bundles to place a robot's eyes where they are most effective.

Magnetic proximity sensors are another possibility, working on the same principle as that described for *tracking*. A basic disadvantage is that the point to be sensed needs to be magnetic. Pneumatic proximity sensors do not have that disadvantage. They can sense every object which interrupts the flow of compressed air from a nozzle or jet pipe, resulting in a back pressure interpreted by a pressure sensor.

Some basic forms of pneumatic proximity sensors are shown in Fig. 9.4.

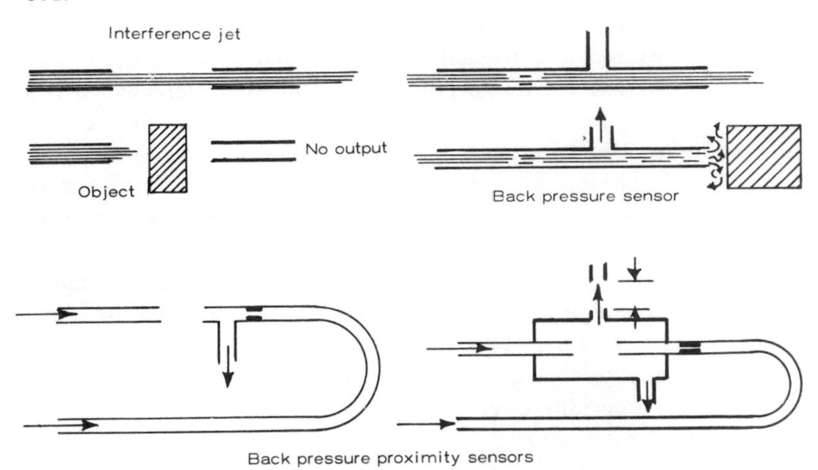

9.4 Some types of pneumatic proximity sensors.

Electrical capacitors can also be used as proximity sensors since their effective capacitance changes slightly when coming close to other objects. Arranged in a suitable AC circuit this can result in a change in signal level sensing the degree of proximity. Such changes, however, will be different for objects in different materials and different sizes, so response will tend to be selective (i.e. most pronounced in proximity to certain objects). Capacitor type sensors, in fact, are rather more usefully employed as *tactile* sensors, working in the same way as touch-sensitive switches.

Tactile Sensors

Tactile sensors, i.e. sensors responding to actual contact or touch, again fall into two broad classifications – those which merely sense contact, and those which not only sense contact but also the degree of contact (e.g. the amount of grip produced by a robot hand). The latter are the more important type — and the more complicated.

Simple touch sensors include capacitors working as a touch-sensitive switch, and mechanical 'feelers', similarly operating a microswitch on contact. Simplest of all is the mechanical stop which brings movement to a halt by a part of the movement running up against the stop.

Tactile sensors which incorporate a sense of grip or feel are usually *force* sensors. The degree of contact or grip is measured as a force which generates a control signal proportional to that force. The control circuit can then be preset so that once the force has risen to the required or maximum safe level, movement stops and there is no further increase in grip.

Force sensors employ a *transducer* transforming force or pressure into electrical signals. Various types of transducers may be used, according to the force levels involved. *Piezo-electric* transducers are suitable for a wide range of forces with excellent sensitivity and signal response. *Carbon graphite* force sensors — working on the same principle as a carbon graphite microphone — are another type, but less sensitive. *Strain gauges* are a further type with a wide range of possibilities, especially where high force levels are involved. Two or more strain gauges arranged in a bridge circuit can also be used to measure *resultant* force where more than one force is involved in the tactile function. For example, gripping force when holding something and the force resulting from turning it at the same time could be determined separately, or as a resultant force on the object involved.

Colour Sensors

Colour sensors employ a rather more sophisticated sort of photo-electric cell which is sensitive to different bands of light. This is used with different colour filters to give maximum signal current when illuminated by a particular coloured light. It could, for example, be arranged to respond or detect only red coloured objects and ignore all other colours.

Equally this basic principle can be developed, with suitable filters and circuity, to extend colour sensing to colour analysis, working in the same way as the colour analysis used in printing colour photographs.

Sense of Smell

As yet there are no practical systems developed to give a robot a sense of smell, although this has already been done in laboratory systems. The nearest off-the-shelf 'smell' sensors which are readily applicable to robots are gas and smoke detectors.

Gas detectors, produced mainly for use in boats and caravans, normally employ a platinum or similar sensory element working on a catalytic basis. In the presence of gas fumes, or vapours from petroleum fluids, the element is activated to generate a 'high signal' current triggering an alarm. Since spillage of hazardous vapour is normally likely to concentrate at ground level, a robot given this form of sense of smell would need to have its 'nose' (i.e. the detector) in its 'feet'.

Smoke detectors work in a rather different way, employing two chambers, one sealed and the other open to the atmosphere, with each containing an ionized stream of helium atoms. A detector in each chamber 'counts' the number of charge particles present and confirms them. In an atmosphere containing smoke, some smoke will enter the open chamber, increasing the number of charged particles present and upsetting the balance between the two detectors. Once this difference is large enough, an alarm circuit is triggered.

The sensitivity of smoke-detectors can be extremely high, and some are claimed to be able to distinguish between different types of smoke, e.g. between cigarette smoke and smoke from a burning fire. Thus a robot so fitted could have some justification for claiming a sense of smell. Also since smoke normally rises, this time the robot can have its 'nose' in its 'head'.

SUMMARY OF SENSOR TYPES

Light sensors — are normally based on a photocell system embracing a light source, optical system, photo-electric sensor and the necessary electrical processing circuit. Light sources used depend on the application, but are normally an incandescent lamp, neon lamp, or solid state light-emitting diode. The optical system is used to concentrate the light source on the sensor. Lenses, prisms or mirrors may be used in applications involving straight-line optical paths. Fibre-optic bundles are used for transmitting light around physical obstacles.

The four types of photo-electric sensors used are photo transistors, photo diodes, photo SCRs (silicon controlled rectifiers) and photovoltaic cells. *Photo diodes* operate in a switching mode, giving a high output voltage and do not require signal conditioning. *Photo transistors* exhibit gain in proportion to the amount of light falling on them and can be used as the first stage of amplification. *Photo SCRs* can switch large amounts of power. *Photovoltaic cells* provide large light-sensitive areas with high sensitivity and fast response.

Proximity Sensors — can work on reflected light, sound, airflow (e.g. *see* Fig. 9.5) and the disturbance of a magnetic field, to name the more usual systems. The most common type of proximity sensor is the magnetic pick-up using a cylindrical permanent magnet mounted behind a soft iron pole piece enclosed within a coil. The magnetic flux traversing the coil and pole piece varies with proximity to any ferrous metal object, generating a voltage in the coil proportional to the rate of change of flux.

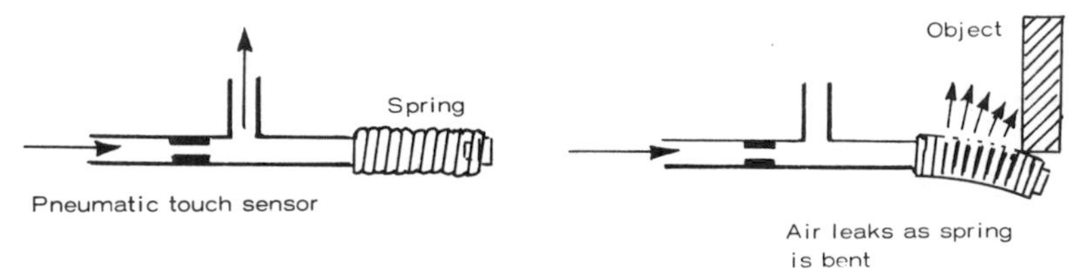

9.5 Pneumatic 'tough-and-deflect' sensor.

A variation on this type is the magnetic sensor which projects a magnetic field produced by an oscillator. This can sense the proximity of any *non-magnetic* metal as a change in load which can be 'read' by a voltage-sensitive network.

Many magnetic type sensors are produced in the form of solid-state proximity switches, having the sensing element and associated electronics combined in a metal cased assembly.

Temperature Sensors — include thermocouples, available in a wide range of physical shapes and mounting arrangements; thermistors; resistance temperature detectors; and the relatively new ceramic temperature sensors. Where it is not possible, or not desirable, for the sensing element to contact the medium whose temperature is being measured, a radiation pyrometer responding to infra-red radiation is used.

Thermocouples work on the principle that when two dissimilar metals are placed in contact a voltage is generated when the junction is heated. This voltage is proportional to the temperature of the junction.

Thermistors work on the principle that with certain semiconductor materials, electrical resistance decreases in proportion to increasing temperature. The change in resistance produced is thus a direct measure of temperature. *Resistance temperature detectors* work the other way round. When heated, the conducting sensing material increases in resistance with increasing temperature, the change in resistance again being a direct measure of temperature.

With a *ceramic temperature sensor* electrical resistance is unaffected until the temperature reaches the Curie point of the material. At the Curie temperature the material undergoes a crystalline change, causing resistance to increase abruptly within a temperature range of less than 5°C. The Curie point, slope and resistance change can be set by using different compositions and treatments of the ceramic material at anywhere from about 60°C to 180°C.

Lasers

A thin laser beam working on the principle of interferometry can provide a near-perfect instrument for measuring both length (distance) and velocity, the short wavelength of laser light providing resolution in the micro-inch range.

Most laser interferometers are based on the Michaelson interferometer in which a beam-splitting mirror is placed at 45 degrees in the laser path. Half of the light beam travels through the beam splitter to a reflecting surface or mirror attached to a distant object, which then returns the beam back to the source. The other half of the beam reflected by the 45 degree mirror is reflected back on itself by another mirror to rejoin the reflected part of the beam returning from the object mirror. Electronic circuitry then counts the number of light-to-dark-to-light transitions, each transition corresponding to a beam movement of one wavelength of light. Using a helium-neon laser which produces red light with a wavelength of approximately 24.5 micro-inches gives a measuring system with a resolution of 3 micro-inches.

Transducers

Strain gauges, variable-capacitance transducers, variable-reluctance transducers and piezo-electric elements are the most common types of sensors used for sensing very small dimensional changes. For larger dimensional changes, potentiometric, inductive and differential-transformer transducers are used.

A *potentiometric transducer* consists of a continuous resistive element fitted with a sliding contact. A fixed voltage is applied to the element and the proportion of voltage which appears at the sliding contact is taken as the output. Force, pressure, acceleration or any similar variable can be used to move the sliding contact, with a proportionate change in output signal. (*See also* Fig. 9.6.)

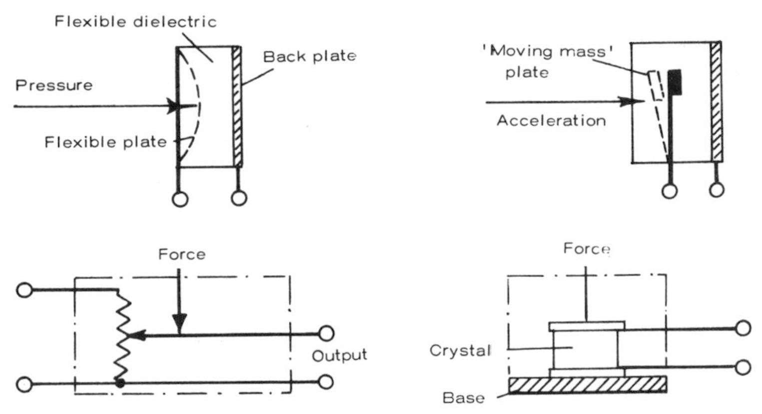

9.6 Examples of transducers — variable capacitance types (top); *potentiometric transducer* (bottom left); *piezo-electric transducer* (bottom right).

A *variable inductance transducer* (Fig. 9.7) converts motion to an electrical signal by movement of an armature or diaphragm relative to a magnetic path, with continuous resolution. They are little used in robotics since they require AC excitation and can produce an

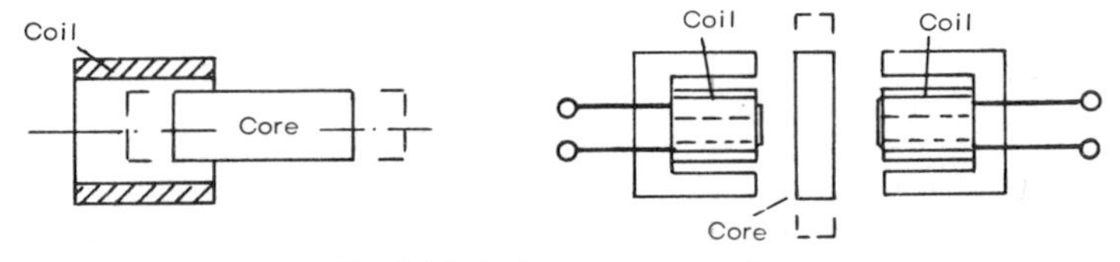

9.7 Variable inductance transducers.

erratic output when mounted in proximity to a magnetic object.

A *differential transformer* can be designed to sense either linear or rotary motion. A movable iron core varies the coupling between the primary and the two secondary windings of the transformer – Fig. 9.8. At the null position output of both secondary windings is the same. As the core moves, inductive coupling will be increased on one secondary winding and decreased on the other winding, providing a differential output. In the case of a differential transformer designed for measuring rotary motion the operating principle is the same but the three windings are mounted on the arms of an E-shaped core and coupling is varied by rotating the iron core.

Again differential transformer transducers require AC excitation, but they provide continuous resolution with high output and low hysteresis. Their main limitation is that they are sensitive to vibration.

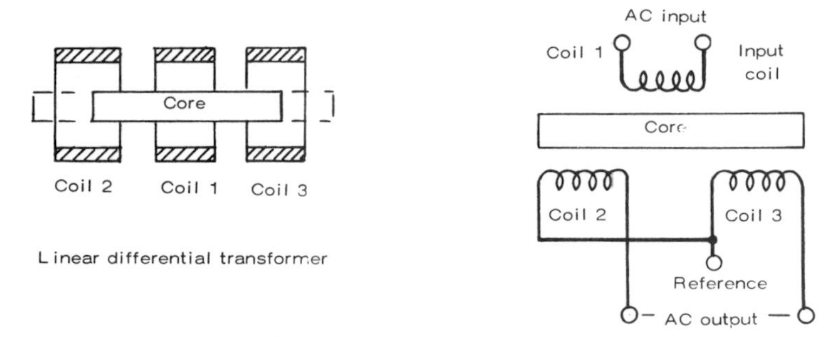

9.8 Linear differential transformer transducer.

CHAPTER 10

The Robot Brain

All the best robots have an electronic or computer-type 'brain'. This can provide a far greater memory capacity in a very much more compact form, and a much greater degree of sophistication in control. It is also much more reliable for there are no contacts to get dirty or become burnt, and no mechanical parts to get 'tired' or wear out. Comparing an electro-mechanical brain with an electronic brain is rather like comparing early computers with modern microprocessors. The first computers used relays and had a bulk which filled a whole room. A modern microprocessor with a similar brain capacity — and far greater reliability — can be no larger in size than a small desk calculator. Yet for all its sophistication an electronic computer or microprocessor still only works on an on/off or 'yes/no' switching basis, using binary digits or 'bits' throughout.

A simplified block diagram of a computer brain is shown in Fig. 10.1. Apart from its *ability* to receive, store and act on information

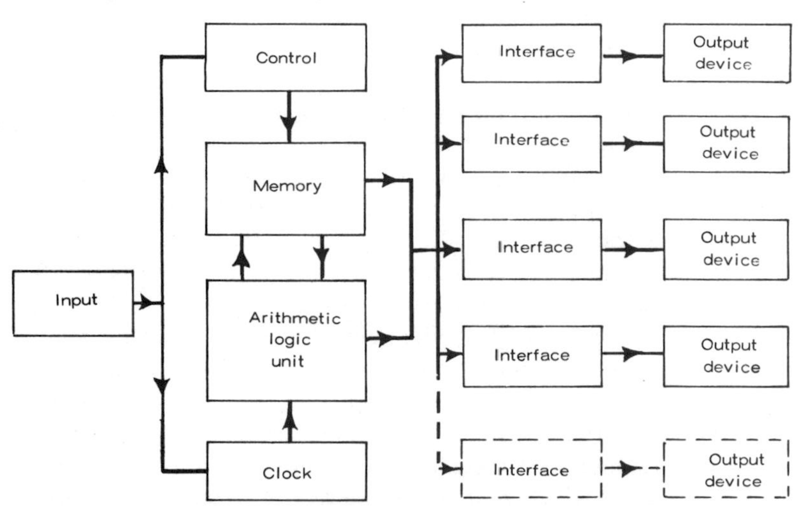

10.1 Block diagram (simplified) of a computer 'brain'.

fed to it, it is a blank brain to start with. It has first to be communicated with in binary language to put a *programme* into the memory where each bit of information is put in its own special place or *address*. Each bit of information stored in this way means 'switch on something' or 'switch off something', the complete programme being fed in and stored in this way. In this respect it is similar to a mechanical programmer (*see* Chapter 11), except that very much more switching information can be stored in a very small space. Alternatively, or if the computer memory is not large enough, data can be stored separately on tape or magnetic discs. (Discs have the advantage of faster recall since an address can be picked out directly rather than having to run through a length of tape.)

The centre of the brain is the control unit. This, in association with an electronic clock, organizes the sequence in which everything will be done, according to the original programme fed in. It calls up numbers from appropriate addresses in the memory and passes them to the logic unit for processing and generation of the correct output signals. These are then interfaced with the appropriate output devices providing the required robot movements, etc.

Programmes can be fed into a computer in the form of punched tape, or more usually nowadays by a keyboard system. This is often linked to a Visual Display Unit (VDU) which allows the operator to see what information is being fed in, and correct if necessary. In *interactive* systems the computer can even query instructions if they are not correct.

The main problem, however, is the language of communication between the human programmer and the computer brain that can only accept instructions in a two-number system, 1 or 0, equivalent to 'yes' or 'no' respectively. To enter decimal number 10, for example, requires the signal 1 0 0 1, or 4 bits; decimal number 100 requires the signal 1 1 0 0 1 0 0, or 7 bits; and decimal number 1,000 requires the signal 1 1 1 1 1 0 1 0 0 0, or 10 bits.

Speed is no problem, for micro-electronics can handle tens of thousands of bits a second, but the process of communicating with a computer in this way in machine code can be slow as far as the operator is concerned, difficult to write, and tedious.

Special computer languages have been developed to overcome this particular problem, where it is possible to give instructions in plain language rather than machine code. Languages of this type used in robotics include Alcol, Fortran, Cobol, Sort, APL, and Assemble; but there are many others. It still takes an operator time to learn such a language, and the various languages used are far from being universal and interchangeable. In fact, a basic requirement for being able to use any language is that a special programme has first been written into the computer through a *compiler* in order for the computer to translate that particular language into machine code – which is the only language *it* can understand. 'Start movement 1', for example, has no meaning at all to a robot brain. It has first to translate this into machine code which then enables it to go to the right place(s) in the memory to determine the appropriate switching instruction(s), determine the logic involved and provide the correct output signal.

A similar sort of problem has been mentioned before. A video camera in place of 'eyes' does not give a robot sight, merely a series of varying electronic signals. It would take an enormous robot brain to analyse and interpret such changing signals in a similar manner to the human brain – and possibly years of work to programme such a brain!

The more complex the problems a robot brain is designed and programmed to handle, in fact, the more exclusive or specialized the robot tends to become. All the many hours of programming, and the full content of the brain, is absorbed in one specialized field. Admittedly the programme could be changed, e.g. if the memory was on tape or disc, merely by replacing the tape or disc with a blank one and starting again. But the *type* of brain devised to suit the original programme might not then be suitable for accepting other types of programmes.

The robot brain can, however, readily be given a degree of self-intelligence to make judgements and act on its own accord. The ability to do this is available in the logic unit. It just needs more information fed into it than is already present in the memory store. This sort of information it can derive from sensors. These can provide additional input signals from 'outside' the programme, which can be compared with original commands (from the memory) and the commands modified accordingly. In other words, the robot brain works according to its internal programmes, but modifies these programmes as and when necessary according to further information received from its own sensors.

A simplified block diagram of a robot brain working on this basis is shown in Fig. 10.2, programmed for five movements: travel backwards or forwards, steering or orientation, arm extension-retraction, and grip. The robot is programmed to move a certain distance in a certain direction, reach out and grasp a delicate orange coloured object from a pile of different coloured objects and carry it safely to another position. Possible hazards in the return path are heavy objects on the floor and a gas flame. These are not always present.

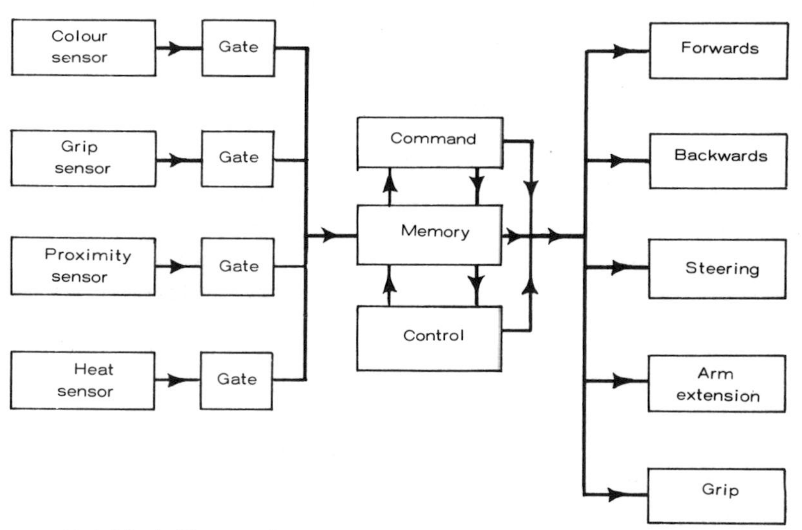

10.2 Block digram of a robot 'brain' with self-intelligence facilities.

The basic programme provides for travel to the right position and arm extension to the position where the objects should be (they are not always exactly in the same place); also a normal return path if there are no hazards in the way. The additional intelligence the robot has to generate from its own sensors, analyse and compare in its own brain are then:

1. Colour sensor signal to identify the object correctly and its actual position; then compare this with the present position of the grip and move the grip to the correct position.
2. Grip sensing, so that the gripper only closes sufficiently to hold and lift the object, not damage it.
3. Proximity sensing so that it can avoid an object in its way on the

return path (modifying the programmed steering motor control).
4. Heat sensor to detect the presence of heat, and again modify the programmed steering motor control.

A little thought will show that some, if not all, of these additional actions could be performed using sensors for feedback control derived from the related output device, not necessarily from additional 'brain power'. The subjects of robot intelligence and sensors are, in fact, very closely related. 'Intelligent' robots are, basically, robots incorporating sensors, so that they can detect changes by themselves outside conditions covered by their programme — and act on such changes by their own decision-making faculty.

Much of the future development of robots and robotics lies in this particular field.

Programmable Controllers

An electronic brain of the type used in robots is correctly called a *programmable controller* (or PC) rather than a computer. Specifically it is defined as a digital operating electronic apparatus which uses a programmable memory for the internal storage of instructions for implementing specific functions. Equally, when a digital computer is used to perform the same function, it too is known as a programmable controller. Mechanical and electromechanical type sequencing controllers (i.e. 'electro-mechanical brains') are *not* regarded as programmable controllers, although they can perform a similar function on a more limited basis.

Unlike most computers, programmable controllers are designed to perform 1-bit combinational logic controls which set and monitor 1-bit operations, rather than process word-oriented information. A *processor* or Central Processing Unit (CPU) is still the heart of a PC, and may be based on discrete components or a micro-processor chip. The former are designed only as 1-bit processors. The microprocessor may have true 'computer' abilities to manipulate word information.

The *memory* of the modern PC is invariably a solid-state device. This is both cheaper and more reliable than a magnetic core memory. It does have one limitation, however. The memory is volatile, i.e. if power is shut off the memory is cancelled (goes blank). Solid-state memory systems therefore need batteries to back up the energy supply in case of power failure. Alternatively a Programmable Read Only Memory (PROM) is used where the information is 'burned into' the memory and cannot be erased. There are also Erasable PROMs (EPROMs) and Reprogrammable Read Only Memories (RPROMs) in which a 'burned in' memory can be erased by means of ultra-violet rays; also the Electrically Alterable ROM (EAROM) which is a Read Only Memory (ROM) which can be modified in-circuit by electrical means.

The maximum memory capacity of a PC system depends on the addressing capability. A typical application for sequential control with few arithmetic, timing or counting functions can be covered by 10–15 instructions, i.e. needs only a 15-bit memory. A more complex industrial robot control system may require some 20–30 instructions to be memorized. Much greater capacity is required when several programmes are to be stored simultaneously, selectable for example by a manually operated mode switch.

Memory systems for PCs are commonly produced in modular form and expandable from a minimum 'starter' kit with a memory of 250 words ($\frac{1}{4}$K) in increments of $\frac{1}{4}$K, 1K and 4K.

Up to 64 counters or *timers* may also be built into a medium scale PC. Time base is normally generated through a quartz oscillator clock which can commonly deliver three different time standards — 1/10 second, 1 second and 1 minute.

Other capabilities or options may also be provided, such as loop controllers (usually in the form of plug-in modules); and Proportional-Integral-Differential (PID), providing a combination of logic and analogue loop control for batch or continuous processing applications.

CHAPTER 11

Electro-Mechanical Brains

Mechanical brains are normally only found in (some) demonstration-type and hobby robots using electrical power. Such a brain comprises a series of timer-operated switches. Each switch is an on/off control for a particular movement or response, either working directly in a low-power circuit or indirectly via a relay (or solid-state switch) in a high-power circuit. The sequence and duration of operation is set up on the timer itself.

The attraction of such a control system is that it is simple to understand, both for designing the programme and seeing how it works. Here, for example, is a desired programme for a walking robot:

1. Move forward for 5 seconds and stop.
2. Pause for 3 seconds, flashing eyes.
3. Turn right through 90 degrees.
4. Pause for 5 seconds and during this period flash eyes.
5. Move forward for 3 seconds and stop.
6. Pause for 1 second, then turn right through 90 degrees.
7. Pause for 1 second and flash eyes.
8. Move forward for 5 seconds and stop.
9. Pause for 1 second, then turn right through 90 degrees.
10. Pause for 1 second, then move forward for 3 seconds and stop.

Hopefully, this should bring the robot back to the starting point — *see* Fig. 11.1.

Now check the time intervals. Are they complete or not? They are not. The programme has not specified any time for 'turning right through 90 degrees'. This can only be determined by a trial run of the robot to see how long it *takes* to turn through 90 degrees. Say this is found to be 2 seconds.

Now add up all the time intervals — 5 seconds forwards, 3 seconds pause, 2 seconds to turn right — and so on — to find the total time for the programme, which in this case works out at 34 seconds.

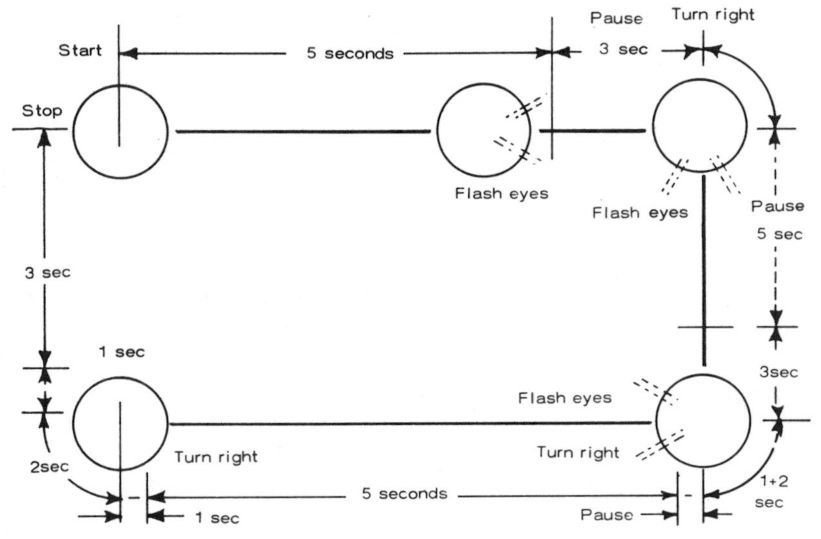

11.1 Planned programme for a mobile robot.

Next step is to identify and separate the different controls or functions involved. There are three:

1. forward/stop – or on/off switching of the drive motor.
2. turn right – or on-off switching of the steering motor with an 'on' time of 2 seconds.
3. flash eyes – or switch on light bulbs (eyes) included in a flasher circuit.

A *timing diagram* for these three circuits can then be drawn up in the form of separate tracks, as shown in Fig. 11.2. Note that these tracks overlap where required, e.g. when the steering motor is commanded to turn the robot right the drive motor must also be switched on for propulsion, otherwise only the steering wheel will move, not the robot.

The three circuits to be switched on/off are shown in Fig. 11.3, bearing in mind that the switches are to be of the simple rubbing contact type which will not carry a high current. The *drive motor* requires a high current (and relatively high voltage), so in this case the command switch is used to complete the circuit to a relay coil. The drive motor current is then carried in a separate circuit and

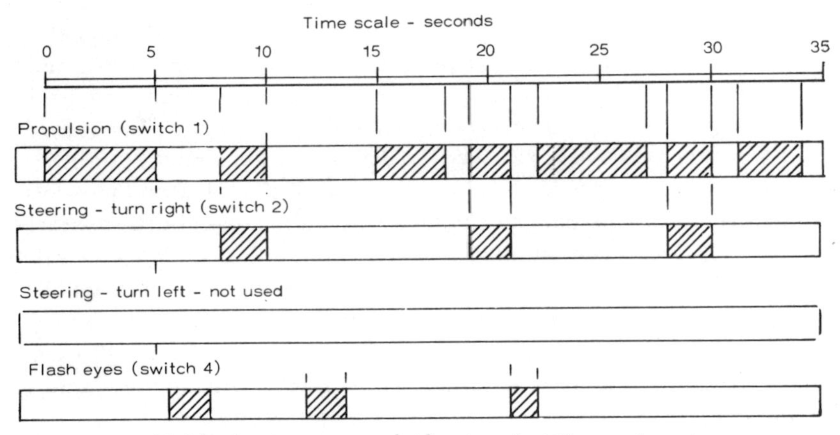

11.2 Robot programme laid out against time scale.

separate battery with the relay contacts working as the on/off switch. Similarly with the *steering motor* circuit. This uses a separate relay and its own separate battery again in the motor circuit, although this could equally well be the same battery as used for the drive motor.

The flashing eye (bulb) circuit does not need a high current. In this case it can use the command switch directly to switch this circuit on/off.

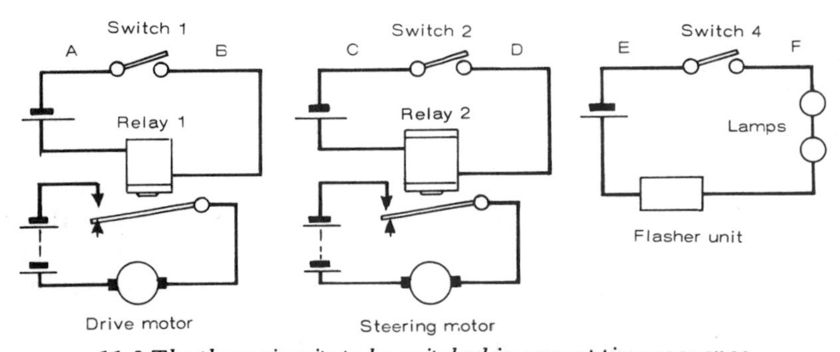

11.3 The three circuits to be switched in correct time sequence.

Translating this into 'hardware', the timing diagram as drawn in Fig. 11.2 could be used directly to duplicate as a linear travel switchboard (e.g. duplicating this in the form of a printed circuit), and traversing the board with a slider carrying four separate brushes. It would then be necessary to arrange a mechanical drive for the slider, timed to travel the length of the board in 24 seconds. (It would also be necessary to arrange that the slider lifts clear of the board and returns to its original position at the end of its travel, ready to repeat the sequence, which would complicate the 'mechanics'; but ignore that for the moment as this is only a demonstration model.)

Circuit wiring would then be as shown in Fig. 11.4. Bridging wires are needed on the back of the board between breaks in the contact strips for circuit continuity, but the overall layout is straightforward. There is also the possibility of eliminating the flasher unit from the eyes circuit. The same effect can be produced by introducing breaks in the appropriate contact strip (bridged across on the back of the board again). This only leaves the 'mechanics' of driving the slider to work out.

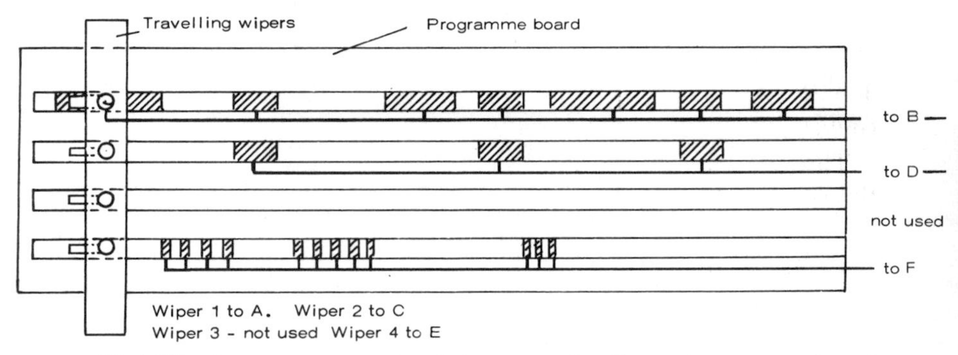

11.4 The programme translated in terms of a linear programme board traversed by wipers.

These can be simplified by using a circular board or disc driven directly by a small electric motor via high reduction gearing so that the time for one revolution is *at least* equal to that of the total sequence time – Fig. 11.5. This will simplify choice of reduction gearing. If the reduction gearing available gives a longer time per revolution, then only that arc of the board corresponding to 34 seconds time is used for the switching circuit. The balance will then be 'return to start' time before the cycle can be repeated.

Contact strips are then plotted as circular arcs on the disc, each aligning with a fixed rubbing brush. In this case the relative patterns are plotted relative to the *position* of each brush – a little more complicated but necessary to get physical spacing between the

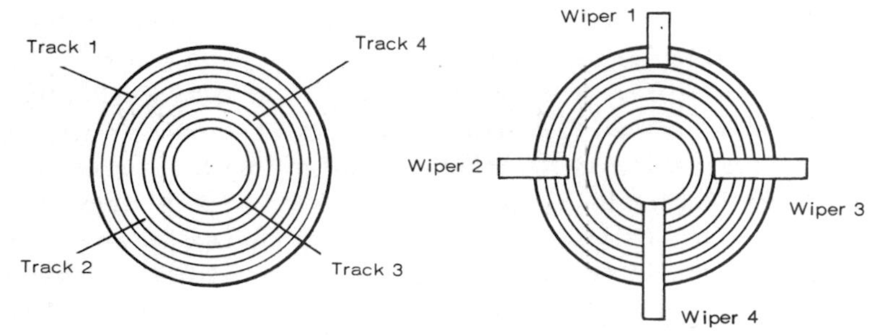

11.5 Disc-type programme boards are more compact, but have less storage area.

brushes. Otherwise connections are exactly the same as with the rectangular board.

This form of master switcher – a rotating disc – provides a straightforward solution where only a few separate commands are required. For a greater number of commands a cylinder driven by a disc is much better, again driven by a small electric motor via reduction gearing at a speed to give one revolution or less in the total sequence time. Contact strips are then stuck on to the surface of the cylinder, each with its separate brush – Fig. 11.6.

Note the greater flexibility of this scheme. Very many more contact strips can be accommodated without overcrowding, merely by increasing the length of the cylinder. Also all the brush contact

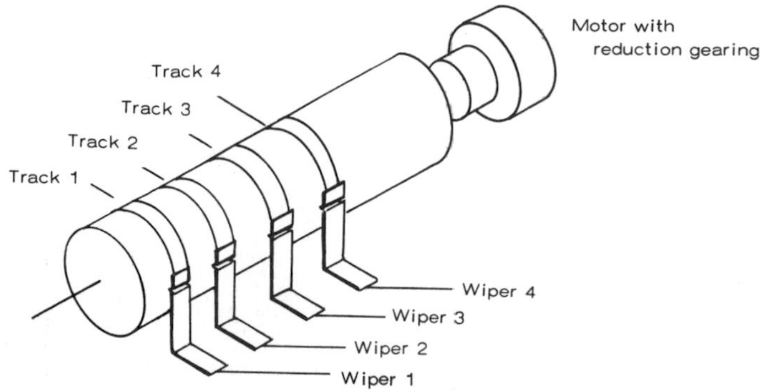

11.6 Cylindrical programmer/timer is the most versatile of electro-mechanical systems.

points are in line, so plotting the positions of the contact strips is simplified; they all relate directly to the positions of the others.

Incidentally, there is an alternative method of using a cylindrical sequence timer, which amateur constructors may prefer. Instead of using contact strips on the cylinder, timing at each circumference line is designated by rounded pegs on similar projections; and brushes are replaced by microswitches of the *changeover* type. The first peg reaching its corresponding switch then switches that circuit on, the next peg reaching the switch after a predetermined interval time then switches that circuit off; and so on.

Such a system makes it much easier to change a programme if required, i.e. adjust the timing. With contact strips you have to adjust timing by lengthening or shortening the strips. With a peg-and-microswitch system you merely have to reposition pegs as necessary. You could, in fact, make the system fully adjustable by drilling a series of holes close together around each circumference line. Pegs are then easily withdrawn and replaced in different positions for adjustment of interval timing.

Whichever method is used, the disc or cylinder represents a complete, individual programme. It is possible to change the programme by changing over to a new disc or cylinder; or have several separate programmes, each with their own timed drive, controlled by a master selector – Fig. 11.7. The master selector could, again, be timer controlled. Alternatively it could be operated manually or by remote control to select the particular programme required.

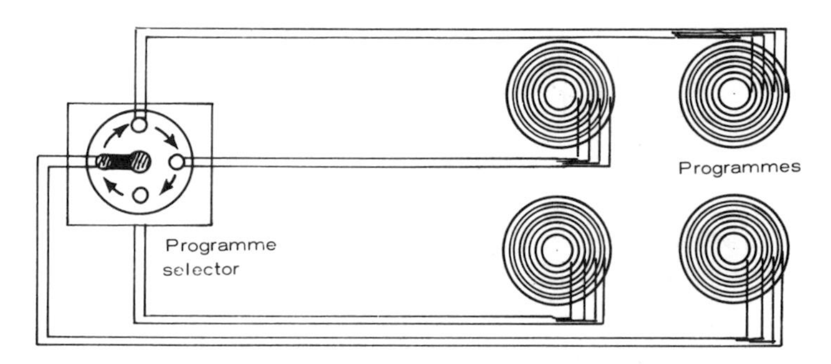

11.7 Separate programmes may be selected by a programme selector.

This describes only the basis of one type of electro-mechanical brain which can be extended, in theory, to accommodate a whole variety of *functions* operating on an on/off basis (even switching on a tape recorder voice, for example) and in *programmed* sequence. Such systems can, however, become quite complicated in the matter of wiring and the bulk of components used, particularly where relays are used to switch higher voltages and currents.

Further problems also arise where motor drives are used which may need to be reversed (e.g. for travelling backwards as well as forwards); or are self-neutralizing (e.g. steering motors). Here again solutions can be provided by relays and limit switches.

Fig. 11.8, for example, shows two circuits for providing forward, stop and reverse control of an electric motor, using either two batteries or a single battery. Either will provide a workable circuit, but a single battery circuit is usually the preferred choice.

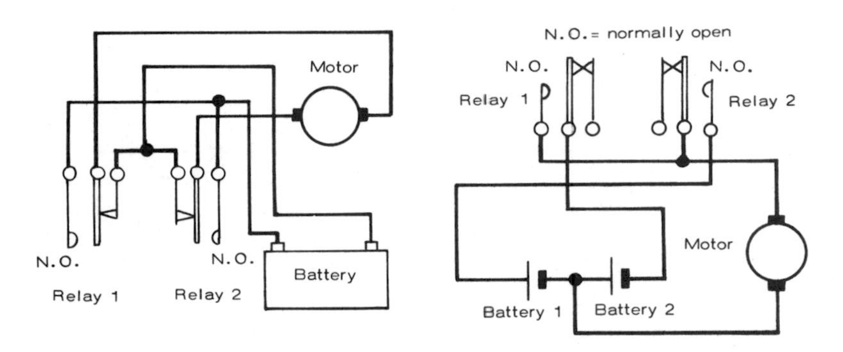

11.8 Two circuits for forward-stop-reverse control of an electric motor.

There are also various ways of making a motor drive stop and switch off automatically when it reaches a certain position. The simplest method is to use limit switches at each end of the desired movement. When the drive reaches these positions the limit switch is operated, breaking the circuit and stopping the motor – Fig. 11.9. Limit switches together with relays can also be used to provide a self-centring action.

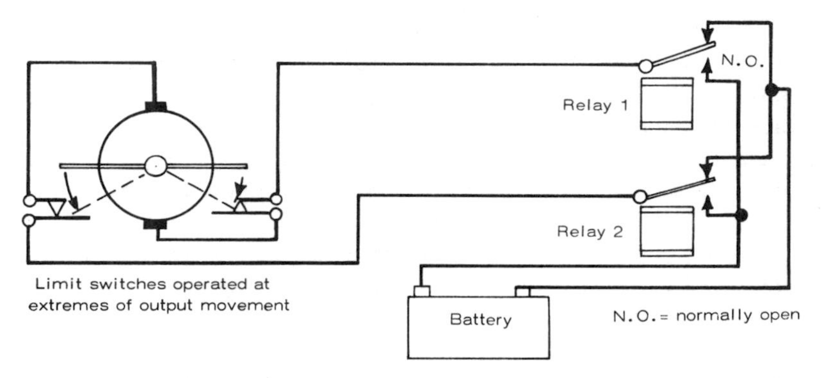

11.9 Circuit for motor control via relays, plus 'switch off' at ends of movement.

Circuitry can be considerably simplified by using solid-state switches instead of relays, together with solid-state power amplifiers where necessary. Fig. 11.10, for example, shows a solid-state equivalent of the relay circuit of Fig. 11.4.

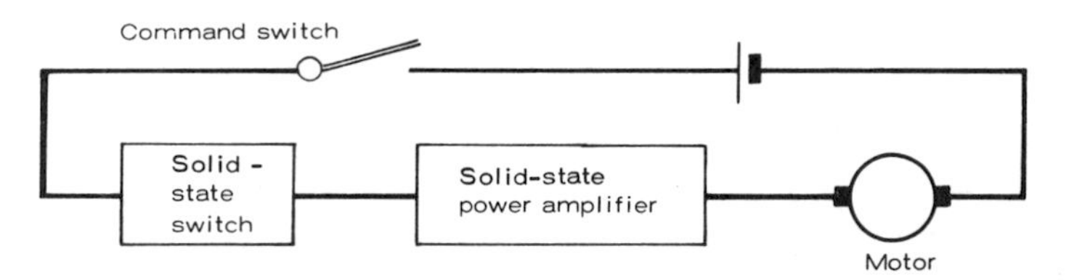

11.10 Solid-state devices are much more compact and generally more reliable than mechanical switches and relays.

CHAPTER 12

Industrial Robots at Work

The main fields of application covered by present-day industrial robots are handling, assembly, hot forging, spot welding and finishing. Robots designed for such work do, in fact, virtually exist as stock items. It would be uneconomic to design and construct individual robots from scratch to perform fairly basic work. So the 'stock' robot for, say, handling, is provided with all the degrees of freedom likely to be required to cover a wide variety of handling jobs; and made available with a variety of wrists and grippers.

Not all robots can be bought and put to use like machine tools, however. There are many fields in which automation is desirable, but off-the-shelf solutions are not available. In such cases robot mechanisms may have to be individually designed and applied in conjunction with microcomputer technology to provide a fully satisfactory, flexible system. Here a robot manufacturer will work in conjunction with the company planning automation to meet required objectives, develop the project in detail, and finally turn it into hardware (the robots) and software (the computer programmes).

Handling Robots

Handling or pick-and-place machines can be called the semi-skilled labourer class of robots. They are particularly useful for handling heavy loads above about 33 pounds (15 kilograms) which is about the maximum a man can handle without mechanical aids. They can perform this duty with exact repeatability, without tiring, and in situations which would be particularly arduous or dangerous to a manual worker. The handling of billets in hot forging and the unloading of diecasting machines are two typical examples of onerous work where a pick-and-place robot can show significant increases in productivity.

Where the movement cycle requires little variation, such robots can be relatively simple designs ranging, say, from a static vertical

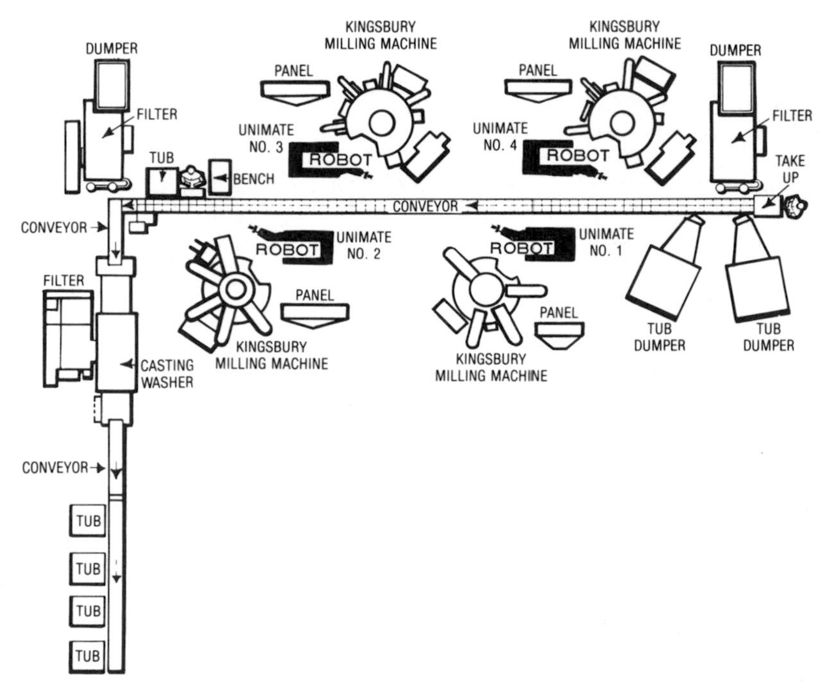

12.1 Typical modern production line using robots for handling and transferring components from stage to stage.

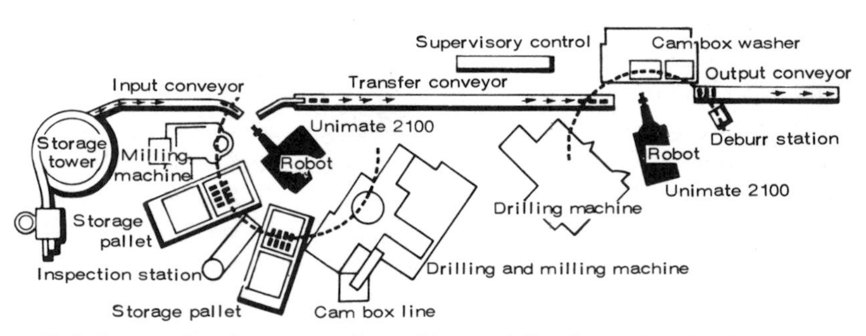

12.2 Conventional automated machine tool line loaded and unloaded by robots.

column with only one degree of freedom to a travelling column with four degrees of freedom, to which a choice of wrist and grippers can be added to suit the particular application. This can avoid the higher cost of multi-axis proportional-controlled robots and still perform the required job, in the same way that it is uneconomic to pay the higher wages of a skilled worker to perform a semi-skilled job. Thus with pick-and-place robots, in particular, robot manufacturers normally offer a choice of different types. The simplest ('least skilled') which will do the job is the logical choice. Equally, some robots will be designed primarily to do a specific, but commonplace job.

A typical example here is the robot designed to feed heavy billets of metal (up to 70 kilograms in weight) directly into a rotary hearth furnace. This can be a basic pick-and-place robot with a special gripper and protection against the high ambient temperature which it will encounter. A typical programme for such a robot would be:

1. pick up the cold billet from the feed conveyor and place it in the furnace (temperature 1500°C).
2. command the furnace to rotate once the billet is in place.
3. pick up the hot billet from the next position in the furnace and place it on the output conveyor.

This cycle is then repeated over and over again, with a typical cycle time of about 15 seconds.

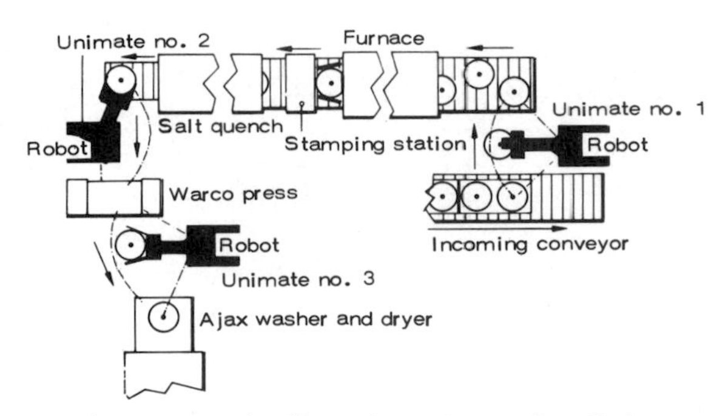

12.3 Automated aus-tempering line using robots to handle hot components. Robot 'fingers' are readily changed to handle different sizes of objects.

Assembly Robots

Assembly robots are normally more sophisticated than pick-and-place robots. Typically, for example, they may have up to 5 degrees of freedom, e.g. three linear movements about one vertical and two horizontal axes, with two rotations for wrist roll and pitch. At the same time displacements may be continuously measured by high resolution optical transducers, giving a resolution of the order of 0.002 mm and accuracy of positioning within 0.1 mm. Such robots can be very energetic workers, and humans must keep well clear of them. Linear movements can accelerate at 4 metres/sec^2 to speeds of the order of 40 metres/min or 90 mph; and in arm-swinging movements the tip of the arm may be travelling at twice this speed!

General Purpose Robots

General purpose robots are generally designed to be available in a variety of forms, starting with a basic machine with 3 degrees of freedom (three axes of movement), increasing to 5 with a two-axis wrist; or 6 with a three-axis wrist. They are thus adaptable to a wide range of common handling requirements of components or materials, or the manipulation of tools and equipment. You start off with a basic robot and add the extra degrees of freedom if you need them; also the type of gripper required for the job.

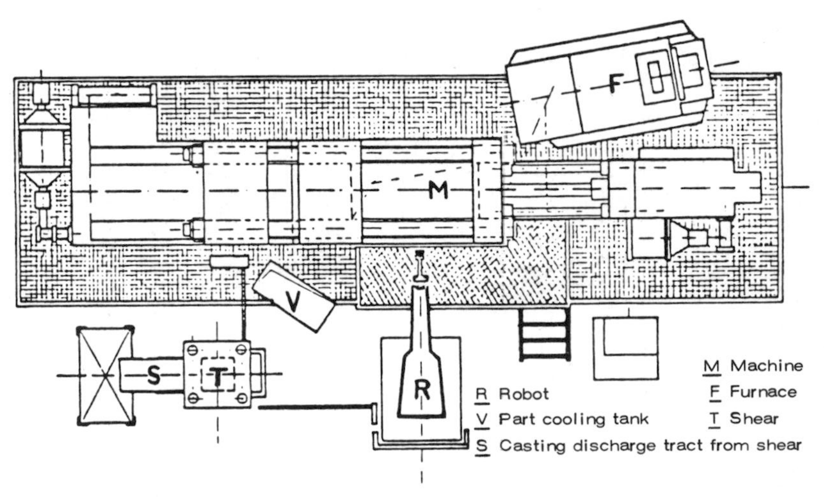

12.4 Arrangement of equipment at Fiat foundry with all the handling work performed by one robot.

Linear movement speeds of the same order as above are typical, with a speed for rotary wrist movements of about 60 degrees per second. Typical gripping power is also interesting and depends on the muscle power used. With electric motor power, typical maximum grip is of the order of 66 pounds (30 kilograms). With pneumatic power up to ten times this grip may be available; and with hydraulic power, up to ten times greater again (probably as much as the total weight of the machine!).

Typical Forging Installation

A complete production line may have several different robots working in conjunction. Fig. 12.6, for example, shows the layout of a typical forging installation. At the start of the line, Robot 1 is a pick-and-place type which selects a billet from those randomly loaded in a bin and lines it up in place on the input conveyor carrying the billet through the induction furnace. Emerging from the furnace, the hot billet is then picked up by Robot 2 and positioned precisely in the first die in the forging press. Components are then handled through the 3-station press by Robot 3 and his twin brother working in exact synchronism. After forming in the third die these twin robots lift the components ready for collection by Robot 4. Robot 4 has two arms, one to handle and transfer the hot component to the trimming press and on to the output conveyor. The second arm comes into operation when the component reaches the trimming press and neatly trims off the flash.

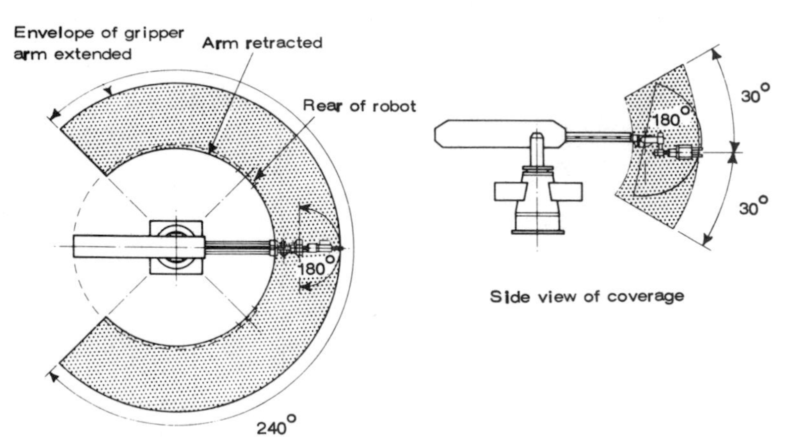

12.5 Typical general purpose transfer robot by Hall Automation showing movements and working areas.

Spot Welding

Spot welding of large assemblies is another job for which robots are well suited. General purpose robots may be used equipped with a welding gun operated in conjunction with an overhead welding transformer; or special welder-robots where the welding transformer is located in the robot itself. The latter avoid the use of long cables with consequent power losses, and the risk of mechanical damage to such cables.

The car industry provides the most outstanding examples of the use of robot spot-welders, and particularly the Fiat Robotgate production line at Turin. This was designed *for* robots with overall computer control rather than a conventional production line with robots replacing human operators. Chassis are delivered from one shop and body panels from another. The panels are loaded into jigs and joined with a few tack welds, then automatically measured and

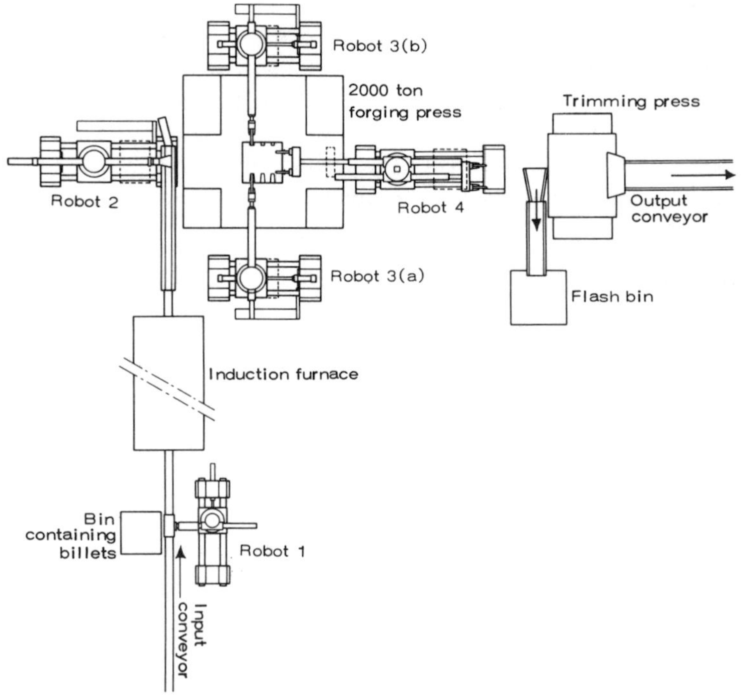

12.6 Layout of a typical forging installation using robots by Fairey Automation.

checked for alignment, as they are lowered on to a chassis which has arrived at the body panel station on its own trolley. It is then passed on to the spot welding stations.

There are several of these as each station only spot welds one area of the body. This provides flexibility. For example, suppose there are six welding stations, where the areas covered by stations 1 and 2 and 5 and 6 are common to all cars passing through the production line, with station 3 welding the appropriate areas on two-door bodies and station 4 the appropriate areas on four-door cars. A two-door car would therefore pass through stations 1, 2, 3, 5 and 6; and a four-door car through stations 1, 2, 4, 5 and 6. The computer identifies the type of body at the start and directs the trolley through appropriate stations following magnetic tracks on the floor.

The computer also does more than this. Welding stations are duplicated so it directs each trolley in turn to the nearest vacant station available in the correct sequence without getting in the way of another trolley travelling out of or into another station. If one station breaks down, it merely directs the trolleys elsewhere to alternative stations to keep the production line flowing. Duplication of stations also means that to accommodate an entirely new model, individual stations can be reprogrammed one by one to accept a flow of new models without shutting down production of present models.

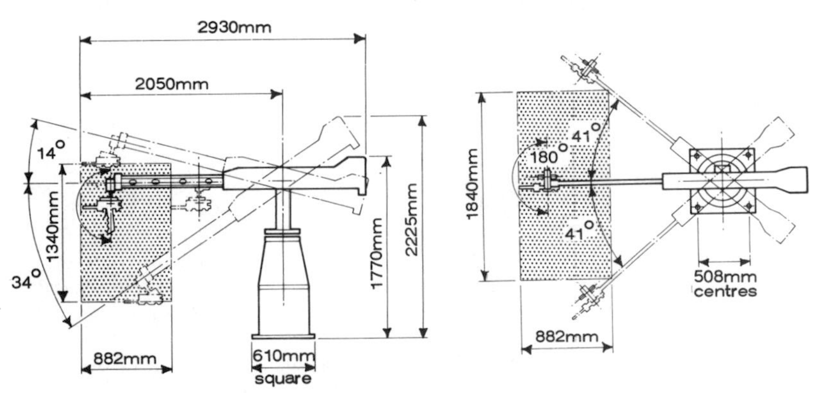

12.7 'HAL' robot welder showing coverage areas.

Seam Welding

Seam welding requires a different technique since the weld is continuous and follows the line of the seam at the same time as the welding torch is oscillated across the seam. Several things can go wrong when an industrial robot is used for this sort of work, programmed on a simple start-and-stop basis. Accurate fixtures are needed to hold and locate the parts being welded, and if these become worn, or the parts are not accurately aligned, the robot welder may start at the wrong place, or the seam path may not coincide with the programmed path. As a consequence the robot welder blindly follows its programme and proceeds to 'weld' a line across a piece of solid metal instead of along the seam.

The more sophisticated robot seam welders incorporate a sensor which can detect the correct starting point and adjust the programme in accordance with the shift in start point. Further, it continues to detect the line of the seam, so that the weld will always follow the correct line. In other words, the sensor modifies and adjusts the original programme, as necessary, to produce a correct weld, regardless of position variations from one part to the next.

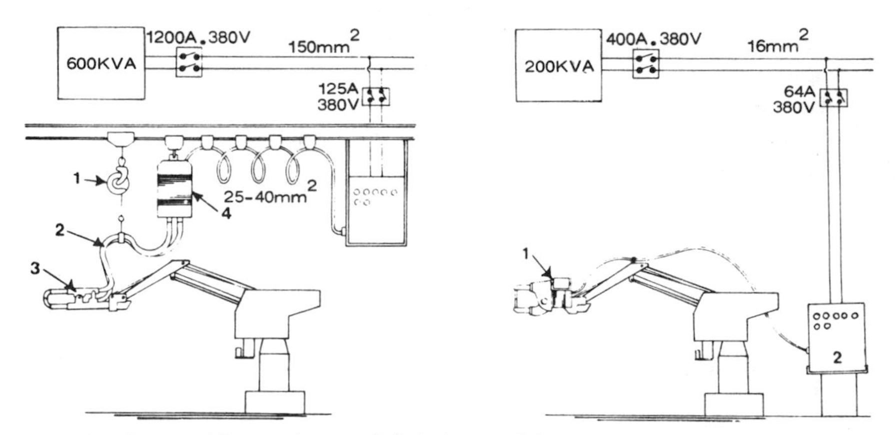

12.8 Spot-welding robot on left is big and heavy requiring balances (1), secondary cable (2), and heavy transmission(4). Integral (Babot) robot right has gun with built-in transmission (1) and separate control box (2).

In Cincinnati's 'Milacron' robot arc welder which works on this principle the sensor actually works through the arc, measuring changes in the arc's current as it oscillates across the seam and sending corrective signals as necessary to the control system. A 'through-the-arc' sensor eliminates the problems which could be associated with external sensors in the severe environment of arc welding, and at the same time keeps the area of the arc clear of any external guidance equipment that could possibly obstruct the passage of the weld gun.

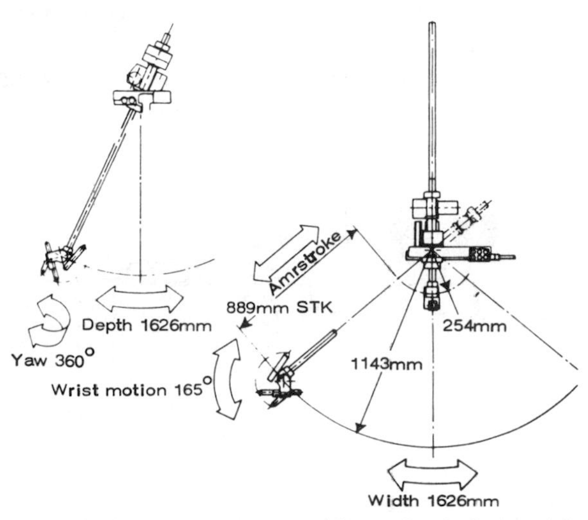

12.9 Unimate 'APPRENTICE' arc-welding robot is designed for portability, the arm weighing only 34 kg. Designed to be used when the workpiece to be welded cannot be moved.

Spray Painting

Undoubtedly the best robots for spray painting are the playback type which are 'taught' the motion programme required by a highly skilled human operator (*see* Chapter 3). To do this it has to have movements which can exactly repeat the complex human movements involved in a spraying sequence. The spray gun is attached to the 'wrist' of the operating arm, which 'bends' at an elbow and can also rotate about a fixed base. The wrist must also be able to rotate in two different planes.

The programme taught by the human operator is retained in the memory of the microcomputer-based control unit; or on a separate tape, together with other necessary information. For example, the computer control system also identifies when the component to be sprayed has arrived on its conveyor line and is correctly in place. Also more than one programme can be held in the computer memory, so that the computer can identify the component by shape from a number of different shapes on the same conveyor line and select the spraying programme accordingly.

There are other possibilities, too, The robot sprayer can be taught to 'lock on' to the component it is spraying and synchronize movements with variations in conveyor speed. It can even be made to 'communicate' with the conveyor by means of interlocking signals to provide correct timing and sequence of operations.

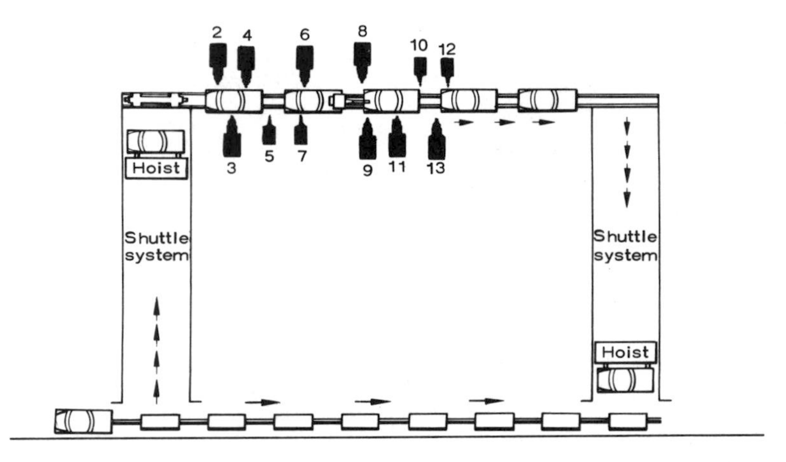

12.10 Spot welding line using robot welders at Chrysler car assembly plant.

Clearly industry in the advanced countries has now reached the robot age, and as already mentioned there are factories in Japan where robots are being made by robots. Equally technological advances in the performance realized by industrial robots will continue to appear, probably more rapidly than most people appreciate. One development of the single-arm robot has already appeared with a capacity to 'change hands' when different tools are needed in an assembly process. Different hands have different tools built into them, to do different jobs; and it changes hands when required by itself. Just as robots can replace manual workers on the production line, a single robot of this type can replace a number of single-arm single-hand robots carrying out a complete assembly process. A type of robot, in fact, which can make other robots redundant! But, even if the multiple-hand robot largely takes over from the single-hand robot, it, too, may be succeeded by second-generation robots with adaptive ability to adjust to changes by itself without re-programming.

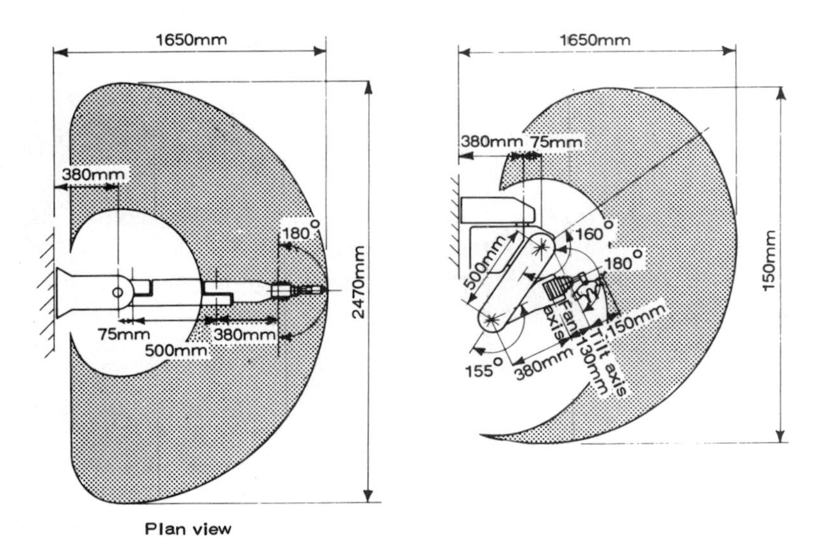

12.11 Coverage of typical playback robot sprayer (Hall Automation 'COMP ARM').

Also in the offing is the 'inspector' robot using both laser and video technology to provide 'vision' needed to handle small parts. This robot can scan parts passed on from the robot production line, check if parts are missing or holes are not drilled where they should be, or drilled incorrectly, and pass or reject the part accordingly. It does this by scanning the part or work and comparing it with a picture of a master model already implanted in its computer-type brain. So even the humans who oversee the work of robots on a production line are not free from the possibility of redundancy. The factory staffed entirely by robots, from Works Manager down to shop-floor level is a practicability already realizable with present technology.

Nobody, in fact, can seriously question the fact that from now on industrial robots will play an increasingly important role, either directly or indirectly, in the lives of us all.

Illustration Acknowledgements

The author and publishers would like to thank the following for permission to use the photographs in the Plates Section:

Battelle, Washington 4; California Institute of Technology 2; Cincinatti Milacron 7, 8; Dainichi-Sykes Robotics Ltd 5, 6; Fairey Automation 9; General Motors, U.S.A. 10, 11, 12 and 15; Peter Holland 3; Novosti News Agency 1.

Index